Introduction to Photonic and Phononic Crystals and Metamaterials

Synthesis Lectures on Materials and Optics

Introduction to Photonic and Phononic Crystals and Metamaterials
Arthur R. McGurn

ISBN: 978-3-031-01256-3   paperback
ISBN: 978-3-031-02384-2   ebook
ISBN: 978-3-031-00248-9   hardcover

DOI 10.1007/978-3-031-02384-2

A Publication in the Springer series
SYNTHESIS LECTURES ON MATERIALS AND OPTICS

Lecture #2
Series ISSN
ISSN pending.

Introduction to Photonic and Phononic Crystals and Metamaterials

Arthur R. McGurn
Western Michigan University

SYNTHESIS LECTURES ON MATERIALS AND OPTICS #2

ABSTRACT

Introduction to Photonic and Phononic Crystals and Metamaterials, by Arthur R. McGurn, presents a study of the fundamental properties of optical and acoustic materials which have been of recent interest in nanoscience and device technology. The level of the presentations is appropriate for advanced undergraduates, beginning graduate students, and researchers not directly involved in the field. References are given to guide the reader to more advanced study in these fields.

Discussions of the physics of photonic and phononic crystals focus on the transmission properties of optical and acoustic radiation arising from their diffractive interaction in these engineered materials. The frequency transmission and non-transmission bands of radiation are explained in terms of the symmetry properties of the photonic and phononic artificial crystal structures. Basic applications of these properties to a variety of their technological applications are examined.

The physics of metamaterials is discussed along with their relationships to the ideas of resonance. Properties of negative index of refraction, perfect lens, and unusual optical effects the new optics of metamaterial media makes available are examined. Related effects in acoustics are also covered.

Basic principles of surface acoustic and electromagnetic waves are explained. These form an introduction to the fundamental ideas of the recently developing fields of plasmonics and surface acoustics.

KEYWORDS

acoustics, electrodynamics, optics, nanoscience, device technology, photonic crystals, phononic crystals, plasmonics

To my wife Maria

Contents

Preface

A recent focus in nanotechnology has been on the study of the properties of engineered materials. These are types of composite media that are designed to exhibit specifically tailored responses when subject to external stimuli. Such media are fashioned by the inclusion, in a background material, of nanoscale structures which complement or augment the composite properties over those directly available in nature. In this way, enhancements in the composite responses yield materials extending the range of technological applications. The new responses of the engineered properties of the composites are evoked by the nature of the inclusions themselves or the pattern in which the inclusions are introduced into a background, supporting, medium.

Important new developments in the areas of engineered media are the design of photonic and phononic metamaterials and crystals. These types of engineered materials allow for the confinement or manipulation of the flow of light and sound in space in ways which have not previously been possible. They have made for the development of technologies based on modulating optical and sound signals reminiscent of the processing of electronic signals in electrical devices. This offers the opportunity to extend, complement, or even replace current electronic device applications through newly available optical and acoustic functioning systems.

Photonic and phononic crystals have designs based on a periodic pattern of optical and acoustic material inclusions in a background supporting medium. Their composite structures are fashioned to respond (in the case of photonic crystals) to optical or (in the case of phononic crystals) to acoustic frequency stimuli that have wavelengths of order of the periodicity of the crystal patterning. At such wavelengths the response of these crystalline materials is governed by a series of frequency pass and stop bands allowing or inhibiting the propagation of the optical or acoustic frequency stimuli through the bulk of the composite material. The band structure properties generated by the periodicity are completely analogous to the electronic band structure properties found in electronic semiconductors. They arise from the diffractive response of the optical or acoustic radiation to the system periodicity.

In electronic semi-conductors the periodic lattice of positive ions in the materials create a frequency band structure of stop and pass bands for electron propagation in the system. In this scheme, only electrons with pass band frequencies can pass through the bulk of the semiconductor. In this regard, the periodicity found in the electronic, photonic, and phononic crystalline systems makes the mathematical treatment and structure of the electron, optical, and acoustic systems exhibit a similarity to one another.

The presence of pass and stop bands in photonic and phononic crystals forms the basis of a number of new technological applications. These range from the design of photonic (phononic) crystal circuits for channeling light (sound), to optical (acoustic) resonator cavities, to the enhancement or suppression of surface radiation. All of these mentioned applications are based on the ability of periodic systems to exclude radiation from their bulk. In addition, the introduction of optical and acoustic nonlinearity to the materials offers for the development of optical and acoustic transistors and rectifiers. These are a great advance in the development of optical and acoustic circuit technology. Advances involving nonlinearity have also been made in the applications of periodic structures in the enhanced generation of higher harmonics of radiation.

Unlike photonic and phononic crystals, metamaterials are designed to interact with wavelengths of light and sound which are much greater than the nearest neighbor separations of the inclusions contained within them. In this limit of wavelengths, the metamaterials appear to act as macroscopic homogeneous materials, and the responses of the systems are refractive responses. The appearance of the homogeneity of the metamaterials in this way is the same as that found in optical crystalline materials which are characterized by a dielectric constant. In the characterization of the optical crystalline and metamaterials by a dielectric constant, the individual interactions of light with each of the atoms or inclusions composing the material is expressed as an average response of the macroscopic medium. Similar average acoustic responses are developed in acoustical metamaterials and are manipulated to display novel macroscopic acoustical responses.

An important new feature in the design of metamaterials, which is used to tailor its response properties, is the introduction of nanoscale inclusions exhibiting frequency resonances. These resonant inclusions in the metamaterial allow the material to provide a variety of novel responses to external stimuli at frequencies near the resonant frequencies of the inclusions. One important new property made available in this way is that of the development of homogeneous media displaying a negative refractive index. Another is the tailoring of designer surfaces which allow for the enhancement of second harmonics generation or the design of new techniques of near field microscopy. The elaboration of ideas based on each of these techniques has formed the basis of a number of developing areas of technological applications in optics and acoustics.

Of all the new technology now available, negative index of refraction has been of particular recent interest. It essentially represents the extension of Snell's law to the consideration of media characterized by negative index of refraction. In this way, the design of materials that can exhibit an extended range of index of refraction has made possible cloaking devices, the development of materials mimicking effects in relativity, etc. These have been the most striking developments in the field of metamaterials but are by no means the only successes found in the applications of these recently developed materials.

To summarize, the responses of photonic and phononic crystals are provided by the pattern of the nanoscale inclusions while in metamaterials the resonant responses of each of the inclusions provides the dominant effect. These two types of media are sources of new diffractive and refractive effects for applications to technology. The focus of this book will be on the elaborations of these ideas and discussion of their consequences.

The book which follows is meant as an introductory text for students to the basic phenomena of the fields of photonic and phononic crystals and metamaterials. It starts with a brief review of basic acoustics and electrodynamics followed by a treatment of optical and acoustic scattering from cylinders. This is important as two-dimensional photonic and phononic crystals are often designed as periodic arrays of cylinders. A chapter is then given which focuses on the properties of two-dimensional photonic and phononic crystals. This is followed by a chapter outlining a treatment of acoustic and optical metamaterials. The last chapter is a discussion of the nanoscience applications of surfaces and layered structures. The focus of the treatments is to introduce the basic ideas of the subject through simple theoretical models with a concentration on analytical models. The student is often referred to the literature where a more advanced development of the subject with more complex models requiring computer solutions is given.

Arthur R. McGurn
Rancho Mirage, California
December 2019

CHAPTER 1

Introduction

Once again, a subject of technological importance is based on formulations involving some of the simplest ideas in physics [1, 2, 4–6]. These are the ideas of resonance [1, 2, 4–19] and periodicity [1, 2, 4, 11, 20–24] which are common to many of the different disciplines of physics. In particular, the applications of these two ideas as focused upon in this book are foundations in the design and engineering of new types of materials exhibiting properties not found in naturally occurring media [7, 11, 21, 22]. On one hand, the idea of resonance occurs in the study of the response of a wide variety of physical problems where it accounts for the generation of enhanced properties in many different types of systems [7, 11]. On the other hand, the idea of periodicity is responsible for the absence of response observed in other types of physical systems [11, 21, 22]. The inhibition and amplification of the responses of a system to external stimuli are important principles in engineering applications, offering a variety of solutions to many different types of technological problems [11].

In this book, the manifestations of the ideas of resonance and periodicity are focused on in the design of metamaterials and photonic and phononic crystals. Metamaterials are engineered materials which are designed to be homogeneous in their applications and to exhibit enhanced properties over those of naturally occurring materials. They derive their important responses to stimuli from the resonance properties included in their designs. Photonic and phononic crystals, on the other hand, are materials which are engineered to display the properties of periodic media in their applications. These include stop and pass band structures representing sets of propagating and non-propagating solutions. From these two types of media it will be seen that a variety of new optical and acoustic properties arise. Many of these are not only of importance in terms of the physics but also afford a variety of new device applications.

In this introduction a qualitative review of the general nature and properties of metamaterials and photonic and phononic crystals are summarized. The remainder of the book then focuses on developing a further quantitative and qualitative understanding of these materials and their responses in applications.

METAMATERIALS

The idea of designed materials based on features of frequency resonant response have developed into the study of the so-called metamaterials [7, 11–19]. These are artificial structured materials which extend the range of physical properties of optical or acoustic media. Metamaterials are composite structures, but the inclusions in the embedding are usually engineered so that the metamaterial composite exhibits frequency responses outside the realm of naturally occurring substances. In this way, the engineered media are designed to act as homogeneous materials in their response to a particular range of external stimuli. Consequently, they are usually created by embedding an array of engineered inclusions into an otherwise homogeneous background medium. The enhancement of the metamaterial frequency response is generally due to a resonant response of the embedded inclusions.

In this regard, an important extension of the physical properties is the development of new media characterized as materials with negative refractive index [7, 11, 25, 26]. Until the development of metamaterials, no such negative indexed media were known or available for engineering applications. The feature of negative index of refraction of metamaterials now extends the ability of optical and acoustic media to change the flow of optical and acoustic radiation through space [7, 11, 26]. It forms the basis of cloaking devices, devices that model relativistic effects on radiation, the design of perfect lenses, and has been introduced as an element of some sensor devices [11]. There are also interesting Doppler effects and properties of Cherenkov radiation arising from the novel properties of negative indexed media [7, 11, 27].

The idea of an embedded structure exhibiting a homogeneous frequency response characterized by an index of refraction is not new to physics [1, 2, 4]. Solids, liquids, and gases are all characterized by index of refraction. Here the characterization is based on considering wavelengths of light or acoustic waves which are much longer than the interatomic or intermolecular separations [1, 2, 4, 7]. This is the realm in which the media gives a refractive response. In the opposite limit of radiation with wavelengths less than the interatomic or intermolecular separations, the responses are diffractive responses. In this latter region the medium no longer appears to be homogeneous, but, rather, the detailed structure of the atomic arrangement becomes important. The diffractive responses in this limit are at the foundations of x-ray diffraction techniques [1, 2, 4]. These diffractive techniques are important in the study of the atomic structure of solids with counterpart acoustical effects [1, 2, 4, 28, 29] involved in the nondestructive testing of composite elastic media.

The ideas involved in the design of metamaterials take the principles of condensed matter physics to the next level of length scales [7, 11]. Metamaterials, formed by embedding resonant structures in a homogeneous background medium, appear to be homogeneous to stimulus radiation with sufficiently large wavelengths. In this limit the resonant features allow the properties of the metamaterial to be tuned at stimulus frequencies near the resonant frequencies of the embedded units [7–19, 25–27]. Consequently, the metamaterial can exhibit a frequency response to an external stimulus that is outside the realm of naturally occurring media.

In acoustic systems two type of resonant structures have been considered [7–17]. These are mechanical resonators which contribute effects described in terms of negative effective masses and mechanical resonators which contribute as negative effective spring constants. In the continuum mechanics view of metamaterials, the mathematics of the elastic limit of the mechanical responses of the metamaterial composed of masses and spring constants are re-characterized in terms of mass densities and Young's or bulk moduli. Consequently, the negative mass density and elastic moduli in the continuum formulation are a direct consequence of the embedded resonant features of the composite forming the metamaterial.

A typical resonant feature of the metamaterial, contributing to a negative effective mass, is a mass shell containing an inner separate attached solid or liquid filling [7–17]. In such a shell structure, resonant modes can be developed in the relative motion of the shell and its inner mass or liquid filling. At resonance there is generally found in such features a vibrational mode in which the shell and its filling exhibit a periodic motion in directions opposite to one another. This is a type of harmonic motion arising from the coupling of the shell with its filling in which the two components forming the resonator beat against each other. When driven by an external periodic force at a frequency near the resonant mode of the shell-filling feature, the elastic response of the feature to the external frequency can appear to be that of a negative effective mass. As a consequence, the elastic medium of the metamaterial exhibits the atypical property that its acceleration is opposite the direction of the applied force it experiences.

It is also possible to introduce features between the masses which modify and contribute to a frequency-dependent spring interaction between the embedded features of the metamaterial [7–17]. In this regard, a one-dimensional chain model will be discussed later in the chapter on metamaterials in which a series of trusses and attached off-chain masses are introduced in order to generate an effective medium with negative effective spring constants. The response of the effective springs to a stretching or contraction force is for the springs to enhance rather than oppose the applied force. This type of negative spring constant behavior is highly frequency dependent and occurs only near the resonant frequency of the resonant truss structure. It appears upon the application to the chain of an external frequency stimulus and its manifestation depends strongly on the driving frequency.

Such negative effective spring constant behaviors, as those found for the chain, are also observed in higher dimensional vibrational systems. They again arise from properly designed embedded resonant features which have continuum limit representations in terms of effective Young's and bulk moduli. Similar extensions are also available in all dimensions for the design of media exhibiting negative mass densities. In this regard, the response of higher dimensional systems is described in terms of an effective medium representation of the metamaterial properties which can including both negative elastic constants and mass densities.

In optical systems resonators that have been embedded in dielectric media forming optical metamaterials are the so-called split ring resonators [11, 18, 19, 25–27]. These are essentially the basic inductor-capacitor LC resonator circuits encountered in elementary electrodynamics. In terms of the inductive, L, and capacitive, C, components of the split ring resonator device, the resonators have the characteristic resonance frequency [30]

$$\omega_0 = \frac{1}{\sqrt{LC}}. \tag{1.1}$$

A crude representation of a split ring resonator would be formed by cutting a gap in a metal ring so that it takes the form of a letter "C" [11, 25–27]. The resulting split ring device is basically representative of a type of LC resonator circuit. In the design the ring part of the circuit is the inductor and the gap in the ring is the capacitor. The combination of both of these elements form a basic LC resonator circuit.

This simple example of the split ring device, however, is an oversimplification made only for pedagogical purposes. In practical applications more complicated variations of the simple cut ring structure are employed in order to overcome various engineering problems. These include modifications dealing with radiative losses and the patterning of the fields generated by the resonators. The basic ideas applied in all of the improved designs of the more sophisticated split ring resonators, nevertheless, remain the same as the simple split ring device described earlier.

When an external frequency-dependent magnetic field is applied perpendicularly to the plane of a split ring, it couples by Faraday's law to the LC oscillator creating a forced harmonic oscillator circuit [11, 25–27]. The ring is now driven by the external field at the field frequency, while generating a magnetic moment from the currents induced in the ring. At frequencies of the external field near the resonant frequency of the ring, given in Eq. (1.1), the response of the system produces an enhanced diamagnetic ring response. This diamagnetic response is similar to that generated by the orbital motions of atomic and molecular electrons when they interact with external magnetic fields [1, 2, 4], but in the case of the split ring it is a much larger response and at resonant frequencies not available in atoms and molecules. In this regard, through the split ring mechanism, a metamaterial composed as an array of embedded split ring resonators can be used to generate an effective medium with new diamagnetic properties not available in naturally occurring materials. As shall be seen these new optical metamaterials have important applications in developing novel optical technologies [11].

NEGATIVE REFRACTIVE INDEX METAMATERIALS

An important application of optical metamaterials with enhanced diamagnetic responses is in the design of artificial materials which display negative index of refraction [11, 25–27]. In optics, negative index of refraction refers to materials which enter into Snell's law,

$$n_i \sin \theta_i = n_r \sin \theta_r, \qquad (1.2)$$

describing the refraction between two optical media at a planar interface, with a negative refractive index. In Eq. (1.2), n_i and n_r are, respectively, the refractive indices of the incident medium and the medium containing the refracted wave. In addition, the ray in the incident medium has an angle of incidence θ_i, and in the refractive medium the refracted ray has an angle of refraction θ_r. (Note that in acoustics, a similar law can be written for the refraction of incident and refracted acoustic rays and this will be discussed in the course of the text.)

There are no naturally occurring optical materials which display negative index of refraction. The reason for this is (as shall be seen in the course of the text) that such materials do not possess frequency regions in which both the permeability and permittivity are negative. The development of the LC resonator-based optical metamaterials, however, allows for formulations of engineered materials which have both negative permeability and permittivity at the same frequency. Similarly, acoustic metamaterials with appropriate resonant properties can be designed to exhibit negative refractive index [11, 25–27] acoustic media. In this regard, the requirements for a negative refractive index in an acoustic medium is that both the mass density and the elastic moduli must simultaneously exhibit negative values.

A significant feature of both homogeneous isotropic optical and acoustic media with negative index of refraction is that the phase and group velocities are oppositely directed. The phase velocity of a planewave mode is a vector pointing in the direction of the modal wavevector, with a magnitude given by the modal frequency divided by the modal wavenumber. It determines the motion of the planes of constant phase of the modes propagating in the medium. The group velocity, on the other hand, is the velocity of energy transport (usually computed in terms of the propagation of a localized pulse) in the medium. As an example, in a positive index media waves described by the planewave form [11]

$$e^{i(kx-\omega t)} \tag{1.3}$$

propagating in the positive x-direction would represent an energy flow in the positive x-direction. In a negative index medium, however, the energy transport of a wave represented by the form in Eq. (1.3) would be in the negative x-direction. The difference between the energy flow and the phase velocity are found to account for many of the unusual refractive properties of negative index media.

It shall be explained in detail later how negative index media increase the range of refractive properties available to optical and acoustic technology. Here some of these aspects are briefly described without going into the detailed physics. In Fig. 1.1, the difference between refraction from a positive index medium into a positive or negative index medium is qualitatively compared. In Fig. 1.1a, a qualitative representation is given of refraction between two positive index media. An incident wave traveling in the second quadrant of the figure is refracted to a refracted wave traveling in the fourth quadrant of the figure. For this system an incident wave in the second quadrant is always refracted only into the fourth quadrant. In Fig. 1.1b, a qualitative representation is given of refraction of a wave traveling in a positive index medium refracted into light traveling in a negative index medium. An incident wave in the second quadrant containing positive index medium now is refracted only into a refracted wave traveling in the third quadrant containing the negative index medium. The refraction trajectories in Figs. 1.1a and 1.1b are fundamentally different from each other and open many new possibilities for the modulation of light flow in optical systems. Negative index materials extend the angles through which light can be refracted in optical systems [11, 25–27].

Consequences of the extended range of refraction include a variety of new effects and devices. These include the design of focusing lens formed as slabs of negative index material [11, 27]. Such types of slab lenses, under the proper conditions, act as so-called perfect

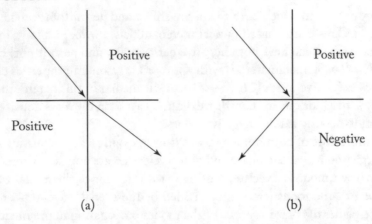

Figure 1.1: Incident and refracted waves at: (a) the interface between two different positive index media and (b) the interface between a positive and negative index medium. The horizontal line is the interface between the two media. In (a) the positive index medium above the horizontal line is different from that below the horizontal line. In both figures the arrows represent the direction of the energy flow.

lenses, faithfully including all possible components of light (i.e., both propagating and evanescent waves) in the images they produce. Unusual new properties are also found in the influence of the new refractive properties on Doppler effects, Cherenkov radiation, antenna applications, Casimir effects, and in the design of cloaking devices [4, 11, 22, 25–27]. In this regard, important potential applications of some of these may include contributions to the design of new sensor and radiating devices [11].

PHOTONIC AND PHONONIC CRYSTALS

Photonic and phononic crystals [4, 11, 20–24], on the other hand, have as a focus the lack of frequency response, respectively, in certain frequency bands of electromagnetic and acoustic waves. Specifically, in the case of photonic crystals electromagnetic radiation propagates through the bulk of the crystal only in so-called frequency pass bands. Outside these pass bands radiation in the so-called stop bands does not pass through the bulk of the crystal. Both the stop and pass bands are continuous segments of the electromagnetic frequency spectrum, uniquely covering the entire spectrum of electromagnetic radiation.

Similarly, phononic crystals also exhibit a series of stop and pass frequency bands but for acoustic radiation rather than electromagnetic radiation. These frequency bands represent the frequencies of acoustic radiation which are excluded from (stop bands) or allowed to pass (pass bands) through the bulk of the phononic crystal. In both photonic and phononic crystals the structure of stop and pass bands arise from the periodic properties of the dielectric or elastic media composing the crystal.

A band structure of stop and pass band frequencies arising from the system periodicity is familiar from the study of semiconductors [1, 2, 4, 11, 20–24]. In the case of semiconductors, the band structure refers to the energy levels of the electrons in the system and arises from the periodicity of the positive ion background in which the electrons travel. The periodicity of the positive ions generates a sequencing of pass and stop frequency bands for the electrons propagating in the bulk of a semiconductor. Electrons propagate through the bulk semiconductor at pass band frequencies but not at stop band frequencies. The nature of the stop and pass band spectrum is determined in large part by the Bragg scattering of the electrons from the periodic lattice of the positive ions.

In the same manner, in photonic crystals a periodic array of dielectric materials develops a frequency band structure of pass and stop bands, as does a periodic array of elastic materials in phononic crystals [1, 2, 4, 11, 20–24]. The band structure in these media arises due to Bragg scattering of electromagnetic or acoustic radiation in the respective systems. In both photonic or phononic crystals this scattering requires that the wavelengths of the radiation be of order or less than the smallest repeat distance of the periodic array. In the opposite limit, at wavelengths much larger than the smallest repeat distance the photonic or phononic crystal would appear to be a homogeneous effective medium, i.e., as a type of metamaterial composite medium.

In the discussion of the electronics of conducting media there are considerations of periodic systems in one-, two-, and three-dimensions. All of these media exhibit electronic band structures which give rise to important physical effects with characteristics related to their dimensionality. Likewise, in photonics and acoustics there can be photonic and phononic crystals arranged as one-, two-, or three-dimensional dielectric or elastic arrays. These provide an electromagnetic or acoustic pass and stop band function on the corresponding level of one-, two-, and three-dimensions, depending on the applications begin considered [11, 20–24]. Examples of such applications are found in laser mirrors (1D photonic crystals), lasers and optical circuits (two-dimensional photonic crystals), and radiators (three-dimensional photonic crystals) [11].

Many of the applications of photonic and phononic crystals arise from the exclusion of stop band radiation from their bulk [11, 20–24]. In this regard, they have found applications in the design of laser cavities and waveguides employed in the development of optical and acoustic circuitry. The designs utilized in these systems often function through the ability of photonic and phononic crystals to limit the losses in their applications over the losses of cavities and waveguides based on designs involving more conventional technologies. In addition, to the blocking of radiation the photonic and phononic crystals have the related abilities to act as frequency filters and frequency on-off switches of radiation [11].

Just as stones in a river can be used to channel the flow of water, photonic and phononic crystals functioning at their stop band frequencies can be used, respectively, to channel the flow electromagnetic or acoustic radiation [4, 7, 11, 20–22]. In this way a system of photonic or phononic crystal circuits may be developed and made to act as a circuitry of electromagnetic or acoustic waveguides. These waveguides are the analogy of wires in an electronic circuit and may similarly be employed to stream light through an optical circuit. In this manner the system of waveguides function as a means to transport the radiation and cause it to interact with a sequence of optical processing devices at various points in the circuit topology. In addition, by the introduction of optical or acoustic nonlinearity to the systems the properties of rectifiers and transistors may be mimicked in optical and acoustic systems. These nonlinear applications then form the basis of a variety of optical or acoustic processing devices meant to match and surpass electronic technologies [4, 11, 22].

GOALS

Photonic and phononic crystals as well as metamaterials have applications as sensors, switches, for the control of radiation properties (both enhancement and suppression of radiation), and in the enhancement of imaging processes. While some of their properties and applications are overlapping in their basic nature, others are uniquely based on enhancements (metamaterials) and others on suppression (crystals) of radiation effects. Some of these possible functions depend on operating the material outside its range of design. As an important example, in some regards photonic and phononic crystals can be used to mimic the properties of negative refractive index media as a sort of compliment to a basic application of metamaterials. In addition, applications of both the metamaterial and crystalline materials are found in a variety of one-, two-, and three-dimensional device geometries to suite a broad range of device applications. These and other properties and characteristics will be briefly reviewed in the following text.

However, the primary focus in the text is to provide an introductory understanding of the basic principles of photonic and phononic crystals and metamaterials with only a little review of applications. This should start the student on the way to a mastery of the basic theory and prepare a bases to understand the most recent literature in these fields. In this sense the text is not a comprehensive study but a development of fundamentals of the subject.

The order of the text is as follows: first a brief review of the basics of acoustics and electromagnetism is presented. The first applications of this review are to a study of some simple scattering problems involving dielectric and elastic cylinders. These treatments are followed by discussions of the basic principles of photonic and phononic crystals. After this photonic and phononic metamaterials are introduced with some applications. The text concludes with introductory discussions of surface waves and excitations in layered coatings.

CHAPTER 2

Review of Acoustics and Electromagnetism

In this chapter a review is given of some of the basic aspects of the continuum theories of elasticity [1, 2, 6, 28] and classical electrodynamics [30, 31]. Both theories were developed prior to the 20th century as classical rather than quantum theories and are based on approaches that ignore the detailed atomic structure of the materials being studied. As such, they are known as continuum theories. They represent some of the earliest endeavors at a mathematical description of natural phenomena on a macroscopic level and are formulated treating the response of materials as involving an average over regions containing many atoms. Both are among the most successful theories of the world, facilitating an understand of most aspects of the macroscopic physics observed in that world. They provide an understanding which ignores the atomic details of a system, dealing with approaches based on phenomenological parameters rather than parameters generated from first principles. This is true particularly in the limit of treatments of linear phenomena founded on a stimulus-response approach. In these treatments, an effect is proportionately related to its stimulus. For this linear limit, the bulk of the macroscopic theory had already existed by the end of the 19th century.

It was only in the 20th century that classical continuum theories were found to fail at explaining the novel features of quantum systems [1, 2]. This prompted the development of quantum physics. In addition, the 20th century also saw the rise of important treatments of nonlinear phenomena. These latter discussions involved treatments of stability, chaos, and various frequency conversions and interaction schemes. All of these efforts still continue, as does a resurgence of novel applications based on the older principles of classical continuum theories. This resurgence of continuum work is particular important in the development of new areas of engineering and technology. Many recent devices and their applications arise from such considerations of continuum limit theories as well as do a variety of proposed applications. In this regard, the bulk of this book will focus on these last recent areas of engineering and technology application of classical continuum model physics [1, 2, 4, 28, 30, 31].

Acoustics or elasticity theory [6] was an early development aimed at understanding the deformation properties of materials as they are subject to forces. In particular a question of acoustics is the nature of the internal macroscopic response of a material that is stimulated by applied and internal mechanical forces. This focus is exclusive from considerations of motions of a body through space as a whole, but emphasizes motions related to the internal properties of the materials forming the body. The focus on the elastic properties of a medium as opposed to its mechanical translations or rotations in space is an important aspect of material science even for considerations at a macroscopic level or for the development of semi-classical treatments of physics. A variety of applications are found in nondestructive testing, ultrasonic techniques, acoustic filtering technologies, solid-state properties, etc. [1, 2, 4, 6, 28].

The focus in the following discussions of elasticity is on defining quantities which describe only the deformations of materials [1, 2, 6, 28]. These quantities are then related to one another in a linear response formulation based on the physical properties characterizing specific materials. An emphasis is on the application of this formulation to understand the nature of acoustic waves in these materials. Applications to developing phononic technologies and phononic metamaterials are discussed later in the book [7–17].

Closely related to the acoustics of solid bodies are continuum treatments of fluid mechanics [1, 2, 4, 6, 28]. The adjustments required in going from an elastic theory of solids to a fluid mechanics treatment of liquids and gases are discuss with a goal of treating acoustic waves in fluids. This allows for the introduction of fluid media as components of phononic devices and the development of phononic metamaterials based on a mixture of solid and fluid media.

In a second treatment, the interaction of electromagnetic fields with various dielectric media will be reviewed [30, 31], and the considerations will be for a classical continuum approach. Again, the medium is considered in an average response to the fields, with the average made over regions contain many atoms. The applied electromagnetic fields modify and are in turn modified by the reaction of the charges and currents within a medium. This interaction acts as the basis of all of the technological applications of materials in optics and electronics. Recent applications considered later in the book are directed toward developments in photonic crystals and metamaterials [4, 7, 11, 21, 22].

The focus in classical electrodynamics will concentrate on media described by a linear response to the fields. Based on these ideas the propagation of electromagnetic waves in various novel engineered materials have recently developed an importance in technology. The novel propagation characteristics arising in artificially designed media have made classical electrodynamics an important, active, and exciting field of study. Quantum effects, while also the object of important developments in recent technologies, are not considered in the text.

The newly developed fields of photonic crystals [4, 11, 21] and photonic metamaterials [4, 7, 11] offer many important types of engineering applications and solutions to technological problems following continuum theory approaches. These extend to designs in antennas, sensors, cloaking devices, systems based on negative refraction, electromagnetic filters, optical circuitry, etc., which are reviewed in later chapters as applications of the theory reviewed here [11].

A review of elasticity and fluid mechanics is now given. This is followed by a review of classical electrodynamics.

2.1   ACOUSTICS

Acoustics is concerned with the propagation of mechanical waves in solids, liquids, or gases [1, 2, 6, 28]. Each of these classes of materials exhibits widely different physical characteristics from the others. For example, solids do not flow while liquids and gases may freely flow through space, and while solids exhibit shears liquids and gases need not. In addition, solids are often relatively incompressible compared to gases and some liquids. In this regard, each of these three different types of media requires its own theoretical approach, with a focus on different essential properties exhibited in each case by the material.

As a result of these differences, in the case of solids the mechanical waves are described by elasticity theory. However, in the case of liquids and gases the continuum approach is based on fluid mechanics. Nevertheless, with fluid mechanics treatments, in the study of liquids some fluid properties are emphasized while in the study of gases others are focused upon. These two approaches for solids and fluids are quite different and, consequently, both continuum theories are treated separately in the following. We first begin with elasticity theory.

2.1.1   ELASTIC THEORY

Elasticity theory relates the deformation of solid bodies from their undeformed states to the actions of applied and internal forces within the bodies [1, 2, 6, 7, 28]. These actions are such that upon the release of the body from the deforming forces, the body returns to its original undeformed state. The processes are then reversible. In the linear formulation of the theory, which is considered here, only small distortions of solids are treated. These linear distortions are characterized by strains, representing small deformations of a body, and which are characterized by a strain tensor. In the elastic limit the strains are proportionate to the small stress forces residing in the body. These stresses, in turn, are expressed by a stress tensor.

In the following, first the strains and stresses of a body are defined, and the general features of the tensors representing these quantities are discussed. This is followed by a treatment of the linear relationships between the two types of tensors, and then by a treatment of elastic waves.

Strains

The small distortions of a material are described by a strain tensor which characterizes and quantifies small changes of shape in the material. These changes are the result of applied and/or internal stresses, and in elasticity theory the object is to define the relationship of the strains and stresses to one another. This is done by developing linear forms expressing the stress obtained from a given strain. To start with, we will focus on developing an appropriate form for the strain tensor. This is done by studying two points of a body as they move under a deformation of the material [1, 4, 6, 7].

In terms of the three-dimensional Cartesian space coordinates with orthogonal unit vectors $\hat{x}_1 = \hat{x}$, $\hat{x}_2 = \hat{y}$, $\hat{x}_3 = \hat{z}$, the position of a point in an undeformed body is given by the position vector

$$\vec{r} = x_1 \hat{x}_1 + x_2 \hat{x}_2 + x_3 \hat{x}_3. \tag{2.1a}$$

As the body is deformed, the coordinate system is fixed in space and the coordinates of points in the material change relative to the fixed coordinate system. As a consequence, after a small distortion of the body, the new position of the point in Eq. (2.1a) becomes

$$\vec{r}' = \vec{r} + \vec{s}\,(\vec{r}) \tag{2.1b}$$

where $\vec{s}\,(\vec{r}) = s_1\,(\vec{r})\,\hat{x}_1 + s_2\,(\vec{r})\,\hat{x}_2 + s_3\,(\vec{r})\,\hat{x}_3$ represents the point displacement from its equilibrium. Notice that $\vec{s}\,(\vec{r})$ in Eq. (2.1b) is labeled by the position $\vec{r}$ of the point in the undeformed material and the resulting mapping is a one-to-one mapping [6, 7].

Next, consider the change in separation of two closely spaced points located at $\vec{r}$ and $\vec{r} + \delta\vec{r}$ in the undeformed body. After a deformation of the material, these two points in the deformed material become

$$\vec{r}' = \vec{r} + \vec{s}\,(\vec{r}) \tag{2.2a}$$

and

$$\vec{r}' + \delta\vec{r}' = \vec{r} + \delta\vec{r} + \vec{s}\,(\vec{r} + \delta\vec{r}) \approx \vec{r} + \delta\vec{r} + \vec{s}\,(\vec{r}) + \delta\vec{r} \cdot \nabla\vec{s}\,(\vec{r})\,, \tag{2.2b}$$

where a Taylor expansion in $\delta\vec{r}$ is made on the far right of Eq. (2.2b). Notice that the coordinate system remains fixed in space, and it is assumed the initial and final separation of the two points is small [7].

Consequently, from the expressions in Eq. (2.2) it follows that the initial and final separations (i.e., $\delta\vec{r}$ and $\delta\vec{r}'$, respectively) of the two points are related by [1, 2, 6, 7, 28]

$$\delta\vec{r}' \approx \delta\vec{r} + \delta\vec{r} \cdot \nabla\vec{s}\,(\vec{r})\,. \tag{2.2c}$$

Here,

$$\nabla\vec{s}\,(\vec{r}) = \begin{vmatrix} \frac{\partial s_1}{\partial x_1} & \frac{\partial s_2}{\partial x_1} & \frac{\partial s_3}{\partial x_1} \\ \frac{\partial s_1}{\partial x_2} & \frac{\partial s_2}{\partial x_2} & \frac{\partial s_3}{\partial x_2} \\ \frac{\partial s_1}{\partial x_3} & \frac{\partial s_2}{\partial x_3} & \frac{\partial s_3}{\partial x_3} \end{vmatrix} \equiv \begin{vmatrix} \epsilon_{11} & \epsilon_{12} & \epsilon_{13} \\ \epsilon_{21} & \epsilon_{22} & \epsilon_{23} \\ \epsilon_{31} & \epsilon_{32} & \epsilon_{33} \end{vmatrix} = \overleftrightarrow{\epsilon}\,(\vec{r}) = \epsilon_{ij}\hat{x}_i\hat{x}_j \tag{2.2d}$$

with the two expressions on the far right of Eq. (2.2d) written in the notation of a second rank tensor. Note that the far right of Eq. (2.2d) expresses the strain tensor in terms of an inner product of unit vectors, $\{\hat{x}_i\}$, and the repeated subscripts i and j indicate that these are summed over $i, j = 1, 2, 3$.

In terms of the tensor notation in Eq. (2.2d), Eq. (2.2c) is expressed in the form

$$\delta\vec{r}' \approx \delta\vec{r} + \delta\vec{r} \cdot \overleftrightarrow{\epsilon}\,(\vec{r}) \tag{2.3}$$

and relates the relative separation of two closely neighbored points in the undeformed solid, $\delta \vec{r}$, to their relative separation in the deformed solid, $\delta \vec{r}'$, through a second rank tensor $\overleftrightarrow{\epsilon}\left(\vec{r}\right)$. The tensor $\overleftrightarrow{\epsilon}\left(\vec{r}\right)$ can now be used to develop a quantitative measure of the deformation of the material from its unstressed state [7].

The tensor, $\overleftrightarrow{\epsilon}\left(\vec{r}\right)$, forms the basis for a natural characterization of the elastic deformations of the solid body, but it falls short of a characterization which only quantifies elastic deformations of the material. In this regard, it is an important point to note that $\overleftrightarrow{\epsilon}\left(\vec{r}\right)$ is found to be invariant under a uniform translation of the entire solid body, i.e., it is zero for $\vec{s}\left(\vec{r}\right)$ a constant. This restricts uniform translations of the media from consideration. However, an additional modification is needed in order to construct from $\overleftrightarrow{\epsilon}\left(\vec{r}\right)$ a tensor which completely characterizes only truly elastic deformations of the material. To provide such a correct tensor, the constructed tensor must also be invariant to rotations of the entire body about an axis of rotation. Such rotations again do not represent deformations of the material. The next object is to construct from $\overleftrightarrow{\epsilon}\left(\vec{r}\right)$ a measure acting exclusively as an indicator of the deformations of an object and which is subject only to the elastic properties of the media forming the object.

To this end $\overleftrightarrow{\epsilon}\left(\vec{r}\right)$ can be rewritten as a sum of symmetric and antisymmetric tensors so that [1, 2, 7]

$$\overleftrightarrow{\epsilon}\left(\vec{r}\right) = \overleftrightarrow{\epsilon_s}\left(\vec{r}\right) + \overleftrightarrow{\epsilon_a}\left(\vec{r}\right), \tag{2.4}$$

where the nine elements of $\overleftrightarrow{\epsilon_s}\left(\vec{r}\right)$ are defined by $\epsilon_{s,ij} = \frac{1}{2}\left(\epsilon_{ij} + \epsilon_{ji}\right)$ and the nine elements of $\overleftrightarrow{\epsilon_a}\left(\vec{r}\right)$ are defined by $\epsilon_{a,ij} = \frac{1}{2}\left(\epsilon_{ij} - \epsilon_{ji}\right)$ for $i, j = 1, 2, 3$. This is a standard result that applies to any second rank tensor. It can be shown that, from this partitioning into symmetric and antisymmetric components, the symmetric component $\overleftrightarrow{\epsilon_s}\left(\vec{r}\right)$ is both invariant to translations and rotations of the entire body in space and about any axes in space. As a consequence, it measures only the deformations of the body.

This can be seen by considering the rotation of the material body about the x_3-axis. A rotation of a point $(x_1, x_2) = (r\cos\theta\,, r\sin\theta\,)$ in the material about the x_3-axis by an infinitesimal $\delta\theta$ takes it into the point

$$\left(x'_1, x'_2\right) = \left(r\cos\left(\theta + \delta\theta\right),\quad r\sin\left(\theta + \delta\theta\right)\,\right) \approx (x_1, x_2) + \delta\theta\left(-x_2, x_1\right). \tag{2.5a}$$

From Eq. (2.4) it follows for the rotation of the body in Eq. (2.5a) that [7]

$$\epsilon_{s,12} = \frac{1}{2}\left(\epsilon_{12} + \epsilon_{21}\right) = 0 \tag{2.5b}$$

and

$$\epsilon_{a,12} = \frac{1}{2}\left(\epsilon_{12} - \epsilon_{21}\right) = \delta\theta. \tag{2.5c}$$

It is seen that the component of the symmetric tensor, $\overset{\leftrightarrow}{\epsilon}_s(\vec{r})$, is insensitive to a bulk rotation of the elastic medium. The antisymmetric components, however, are sensitive to rotations and $\overset{\leftrightarrow}{\epsilon}_a(\vec{r})$ is, consequently, not a good indicator of the deformations of the body. The discussions can be extended to all other axis of rotations in space, including all components of the symmetric and antisymmetric tensors in Eq. (2.4), so that $\overset{\leftrightarrow}{\epsilon}_s(\vec{r})$ is the favored indicator of a deformation.

As a test, another indication of the usefulness of the symmetric tensor as a measure of the distortions of a material is provided by considering the length change between the two points in Eqs. (2.2) and (2.3) under a distortion. From Eq. (2.3) it follows that the relationship [6, 7, 28]

$$\delta \vec{r}' \cdot \delta \vec{r}' \approx \sum_{i=1}^{3} (1 + 2\epsilon_{s,ii})\delta r_i^2 + \sum_{i \neq j=1}^{3} 2\epsilon_{s,ij}\delta r_i \delta r_j \qquad (2.6)$$

expresses the distance of two closely spaced points after an elastic distortion in terms of their separation before the distortion. Here the antisymmetric components in Eq. (2.4) do not enter into the relationships for the separations because the rotations they measure do not affect the separation between the two points under consideration.

It is important to note that the separation of the tensor in Eq. (2.4) into symmetric and antisymmetric parts introduces restrictions upon the elements composing each of these tensors. In the separation, the symmetric part of the tensor, describing the body distortions, consists of six independent components while the antisymmetric part is composed of three independent components. Consequently, of the nine entries of the symmetric tensor only six are independent entries. These restrictions of the symmetric form result in a number of compatibility equations which exist between the entries of the tensor and directly arise from the calculus of several variable.

The resulting compatibility equations include the following identities [6, 7, 28]:

$$\frac{\partial^2 \epsilon_{s,ii}}{\partial x_j^2} + \frac{\partial^2 \epsilon_{s,jj}}{\partial x_i^2} = 2\frac{\partial^2 \epsilon_{s,ij}}{\partial x_i \partial x_j} \qquad (2.7a)$$

for $i \neq j = 1, 2, 3$, and

$$\frac{\partial^2 \epsilon_{s,ii}}{\partial x_j \partial x_k} = \frac{\partial^2 \epsilon_{s,ij}}{\partial x_i \partial x_k} + \frac{\partial^2 \epsilon_{s,ik}}{\partial x_i \partial x_j} - \frac{\partial^2 \epsilon_{s,jk}}{\partial x_i^2} \qquad (2.7b)$$

for $i \neq j \neq k = 1, 2, 3$.

From the above discussion the symmetric strain tensor in Eq. (2.4) is seen to be an appropriate representation of the deformation properties of materials. In the course of the discussions the notation used to describe the symmetric tensor has become increasingly complicated. Now that the point has been made a simplification of the notation is in course.

To summarize and abbreviate the notation for the symmetric tensor measuring the strains in a continuum material, we define a new notation. The symmetric strain tensor is represented as $\overleftrightarrow{e}(\vec{r})$ and is obtained from the symmetric tensor in Eq. (2.4) in the following notational change [7]:

$$\overleftrightarrow{e}(\vec{r}) = \begin{vmatrix} e_1 & \frac{1}{2}e_4 & \frac{1}{2}e_5 \\ \frac{1}{2}e_4 & e_2 & \frac{1}{2}e_6 \\ \frac{1}{2}e_5 & \frac{1}{2}e_6 & e_3 \end{vmatrix}, \tag{2.8}$$

where $e_1 = \epsilon_{s,11}$, $e_2 = \epsilon_{s,22}$, $e_3 = \epsilon_{s,33}$, $e_4 = 2\epsilon_{s,12}$, $e_5 = 2\epsilon_{s,13}$, and $e_6 = 2\epsilon_{s,23}$. Here the six independent components of the symmetric strain are traced back through Eqs. (2.2d) and (2.4) to the original linear strain presented in Eqs. (2.2c) and (2.3). The strains are small linear strains so that $|e_j| \ll 1$ for $j = 1, 2, 3, 4, 5, 6$. At larger strains a nonlinear elastic theory must be introduced, and this will not be treated here.

The strain components in Eq. (2.8) will now be related to the stress forces in a solid applying a linear, Hook's Law, type of theory. This considers the leading order effects in the limit of small deformations for which nonlinear effects are absent.

Stresses

The next factor entering into the description of the elastic properties of a solid involves the introduction of the forces which act to create or respond to distortions of its material. In the following, the theory of these stress forces will be developed, with the emphasis put on the internal forces arising from the interaction of the material with itself. The internal stresses will be shown to be naturally described by a symmetric second order tensor. External forces due to interactions with gravity or other external fields will not be considered here but can be introduced independently of the internal stress tensor [1, 2, 6, 7, 28].

To determine an appropriate description of the internal stresses in a material, consider an elastic medium described by a mass density $\rho(\vec{r})$ and an infinitesimal cubic volume of the medium located between

$$x_{10} \leq x_1 \leq x_{10} + \Delta x_1, \tag{2.9a}$$

$$x_{20} \leq x_2 \leq x_{20} + \Delta x_2, \tag{2.9b}$$

and

$$x_{30} \leq x_3 \leq x_{30} + \Delta x_3. \tag{2.9c}$$

This small element of the material has a mass given by $\Delta m = \rho(\vec{r}) \Delta x_1 \Delta x_2 \Delta x_3$, where $\rho(\vec{r})$ is the mass density at the small element, and experiences forces from the material surrounding it. These internal forces can determine the properties of elastic waves traveling within the material [6, 7, 28].

The force at any point in the medium is described by a position dependent vector $\vec{F}(\vec{r})$ which can exhibit a variation over the volume of the medium being treated. In particular, consider the forces experienced by the element in Eq. (2.9) arising from the internal interaction of the medium with itself. (The interaction of the medium with external fields is not an interest here but can be easily introduced into the following treatment.) These internal force contributions are absent in the absence of strains in the medium.

The force from the internal interactions of the medium varies over the infinitesimal volume. Consequently, between the faces of the infinitesimal volume in Eq. (2.9), the change in force between the faces of the volume is given by [6, 7, 28]

$$d\vec{F}(\vec{r}) = \frac{\partial \vec{F}}{\partial x_1}\Delta x_1 + \frac{\partial \vec{F}}{\partial x_2}\Delta x_2 + \frac{\partial \vec{F}}{\partial x_3}\Delta x_3. \tag{2.10}$$

In the absence of bulk forces arising for example from gravitation fields, this variation of the internal force of the medium over the mass of the infinitesimal volume element causes an acceleration of the mass in the volume element. This interaction is described by [6, 7, 28]

$$\Delta m \vec{a} = \rho(\vec{r})\,\Delta x_1 \Delta x_2 \Delta x_3 \vec{a} = d\vec{F}(\vec{r}) \tag{2.11}$$

and relates the acceleration, $\vec{a} = \ddot{\vec{r}}$, of the mass element Δm at $\vec{r}$ to the force differences acting between the faces of that volume element.

Combining the results from Eqs. (2.10) and (2.11) it then follows that

$$\rho(\vec{r})\,\vec{a} = \frac{1}{\Delta x_2 \Delta x_3}\frac{\partial \vec{F}}{\partial x_1} + \frac{1}{\Delta x_1 \Delta x_3}\frac{\partial \vec{F}}{\partial x_2} + \frac{1}{\Delta x_1 \Delta x_2}\frac{\partial \vec{F}}{\partial x_3}. \tag{2.12}$$

An interesting point regarding this expression is that, for our considerations in the following, contributions to $\vec{F}(\vec{r})$ from interaction of the material with external fields would tend to be constant over the material, i.e., such external fields are usually slowly varying over the materials treated. Consequently, for these interactions $d\vec{F}(\vec{r}) = 0$ and only internal forces of the material give non-zero contributions to the $d\vec{F}(\vec{r})$.

Some simplification of the right-hand side of Eq. (2.12), making the expression more transparent, is possible. In component form the right-hand side of Eq. (2.12) can be rewritten as [1, 2, 6, 7, 28]

$$\begin{aligned}
&\left(\frac{\partial}{\partial x_1}\frac{F_1}{\Delta x_2 \Delta x_3} + \frac{\partial}{\partial x_2}\frac{F_1}{\Delta x_1 \Delta x_3} + \frac{\partial}{\partial x_3}\frac{F_1}{\Delta x_1 \Delta x_2} \right)\hat{x}_1 \\
&+\left(\frac{\partial}{\partial x_1}\frac{F_2}{\Delta x_2 \Delta x_3} + \frac{\partial}{\partial x_2}\frac{F_2}{\Delta x_1 \Delta x_3} + \frac{\partial}{\partial x_3}\frac{F_2}{\Delta x_1 \Delta x_2} \right)\hat{x}_2 \\
&+\left(\frac{\partial}{\partial x_1}\frac{F_3}{\Delta x_2 \Delta x_3} + \frac{\partial}{\partial x_2}\frac{F_3}{\Delta x_1 \Delta x_3} + \frac{\partial}{\partial x_3}\frac{F_3}{\Delta x_1 \Delta x_2} \right)\hat{x}_3,
\end{aligned} \tag{2.13}$$

where $\frac{F_i}{\Delta x_j \Delta x_k}$ is the ith component of force per area in the j–k plane. This explicitly expresses the components of the vector in Eq. (2.12) in terms of its components of force per area or stress.

From Eq. (2.13) one can define the elements of a three by three stress matrix composed of the elements [6, 7, 28]

$$\sigma_{11} = \frac{F_1}{\Delta x_2 \Delta x_3}, \quad \sigma_{21} = \frac{F_1}{\Delta x_1 \Delta x_3}, \quad \sigma_{31} = \frac{F_1}{\Delta x_1 \Delta x_2}$$

$$\sigma_{12} = \frac{F_2}{\Delta x_2 \Delta x_3}, \quad \sigma_{22} = \frac{F_2}{\Delta x_1 \Delta x_3}, \quad \sigma_{32} = \frac{F_2}{\Delta x_1 \Delta x_2} \tag{2.14}$$

$$\sigma_{13} = \frac{F_3}{\Delta x_2 \Delta x_3}, \quad \sigma_{23} = \frac{F_3}{\Delta x_1 \Delta x_3}, \quad \sigma_{33} = \frac{F_3}{\Delta x_1 \Delta x_2}$$

so that in matrix format

$$\overset{\leftrightarrow}{\sigma}(\vec{r}) = \begin{vmatrix} \sigma_{11} & \sigma_{12} & \sigma_{13} \\ \sigma_{21} & \sigma_{22} & \sigma_{23} \\ \sigma_{31} & \sigma_{32} & \sigma_{33} \end{vmatrix} = \sigma_{ij} \hat{x}_i \hat{x}_j \tag{2.15}$$

defines what is known as the stress tensor. Note that the far right of Eq. (2.15) expresses the stress tensor in terms of an inner product of unit vectors, $\{\hat{x}_i\}$, and the repeated subscripts i and j indicate that these are summed over $i, j = 1, 2, 3$.

Applying Eqs. (2.13)–(2.15) in Eq. (2.12) it follows [6, 7, 28] that for the volume element at $\vec{r}$

$$\rho(\vec{r})\vec{a} = \left[\frac{\partial \sigma_{11}}{\partial x_1} + \frac{\partial \sigma_{21}}{\partial x_2} + \frac{\partial \sigma_{31}}{\partial x_3} \right] \hat{x}_1$$

$$+ \left[\frac{\partial \sigma_{12}}{\partial x_1} + \frac{\partial \sigma_{22}}{\partial x_2} + \frac{\partial \sigma_{32}}{\partial x_3} \right] \hat{x}_2 \tag{2.16}$$

$$+ \left[\frac{\partial \sigma_{13}}{\partial x_1} + \frac{\partial \sigma_{23}}{\partial x_2} + \frac{\partial \sigma_{33}}{\partial x_3} \right] \hat{x}_3$$

$$= \nabla \cdot \overset{\leftrightarrow}{\sigma}(\vec{r}).$$

The acceleration is then related to the divergence of the stress tensor at each point of the material.

Some consideration of the stress tensor in Eq. (2.15) will indicate that it is a symmetric tensor. This is required so there is no net torque on the infinitesimal volume elements composing the solid, and for the applications discussed in this text this will be the case for all considerations. For example, in order that there be no net torque in the solid about the x_3-axis it follows that [6, 7, 28]

$$\sigma_{12} = \sigma_{21}. \tag{2.17}$$

To see this, consider a cube of material with sides dx_1, dx_2, dx_3 which are of equal length. For a rotation of the cube about the x_3-axis, take the x_3-axis along an edge of the cube passing through the point $(x_1, x_2) = (0, 0)$. To leading order the force in the $\hat{x}_2$ direction acting in the x_2–x_3 plane containing the point $(x_1, x_2) = (dx_1, 0)$ is

$$\vec{f}_2 = \sigma_{12} dx_2 dx_3 \hat{x}_2. \tag{2.18a}$$

Similarly, to leading order the force in the $\hat{x}_1$ direction acting in the x_1–x_3 plane containing the point $(x_1, x_2) = (0, dx_2)$ is

$$\vec{f}_1 = \sigma_{21} dx_1 dx_3 \hat{x}_1. \tag{2.18b}$$

The torque about the x_3-axis acting at the point $(x_1, x_2) = (dx_1, dx_2)$ in the x_1–x_2 plane is then given by

$$\vec{\tau}_3 = (\sigma_{21} - \sigma_{21}) \, dx_1 dx_2 dx_3 \hat{x}_3. \tag{2.18c}$$

The requirement of a zero torque about the x_3-axis is seen to be that $\sigma_{12} = \sigma_{21}$. The symmetry of the other components of the stress tensor follow similarly.

The resulting symmetric stress tensor can be written in a notation similar to that of the strain tenor format in Eq. (2.8) as [6, 7, 28]

$$\overset{\leftrightarrow}{\sigma}(\vec{r}) = \begin{vmatrix} \sigma_1 & \sigma_4 & \sigma_5 \\ \sigma_4 & \sigma_2 & \sigma_6 \\ \sigma_5 & \sigma_6 & \sigma_3 \end{vmatrix}, \tag{2.19}$$

where now $\sigma_1 = \sigma_{11}$, $\sigma_2 = \sigma_{22}$, $\sigma_3 = \sigma_{33}$, $\sigma_4 = \sigma_{12}$, $\sigma_5 = \sigma_{13}$, and $\sigma_6 = \sigma_{23}$.

Linear Relation Between the Stress and Strain Tensors

The two tensors in Eqs. (2.8) and (2.19) are related to one another by a set of linear equations. This is the leading order of the relationship between the stress and strain and comprises the study of linear elasticity theory. It ignores higher-order nonlinear terms which are considered in the study of nonlinear acoustic phenomena. While in linear elasticity there are modal solutions with distinct frequencies, in nonlinear acoustics the solutions involve more complex time dependences.

The linear relationship expressed in component form gives [1, 2, 6, 7, 28]

$$\sigma_i = \sum_{j=1}^{6} C_{ij} e_j, \tag{2.20}$$

where C_{ij} are the elastic stiffness constants which characterize the elastic properties of the medium. Here the elastic stiffness constants can be introduced as phenomenological constants or are derived from first principle studies.

The elastic energy density, characterizing the energy required to make an elastic deformation of the material, is obtained from Eq. (2.20) through a consideration of the work needed to deform the material. The result for the elastic energy density U from such a consideration is given in the linear limit by [6, 7, 28]

$$U = \frac{1}{2} \sum_{i=1}^{6} \sum_{j=1}^{6} C_{ij} e_i e_j. \tag{2.21}$$

This is reminiscent of the result for the work needed to deform a spring, and, similarly, the stress is related to this energy by

$$\sigma_i = \frac{\partial U}{\partial e_i}. \tag{2.22}$$

As an example of Eq. (2.20), consider first an isotropic homogeneous material. This is a uniform medium with complete rotational and translational symmetry. In this limit Eq. (2.20) takes the form [6, 7, 28]

$$\sigma_i = 2G e_i + \lambda e \quad i = 1, 2, 3, \tag{2.23a}$$

where $e = e_1 + e_2 + e_3$, and

$$\sigma_i = G e_i \quad i = 4, 5, 6, \tag{2.23b}$$

where λ and G are the Lame elastic constants. The energy of elastic deformation is then from Eq. (2.21) given by [6, 7, 28]

$$U = \frac{1}{2} (2G + \lambda) \left(e_1^2 + e_2^2 + e_3^2 \right) + \frac{1}{2} G \left(e_4^2 + e_5^2 + e_6^2 \right)$$
$$+ \lambda \left(e_1 e_2 + e_1 e_3 + e_2 e_3 \right) \tag{2.24}$$

and Eqs. (2.23) and (2.24) are related though Eq. (2.22).

In a system with cubic symmetry there are more restrictions on the elastic constants than in the isotropic homogeneous material. For the case of cubic symmetry [7, 28] with cube faces in the x_1–x_2, x_1–x_3, and x_2–x_3 planes, the only nonzero values of C_{ij} are $C_{11} = C_{22} = C_{33}$, $C_{44} = C_{55} = C_{66}$, and $C_{12} = C_{21}$. With these considerations, Eq. (2.20) takes the form

$$\sigma_1 = C_{11} e_1 + C_{12} (e_2 + e_3), \tag{2.25a}$$
$$\sigma_2 = C_{11} e_2 + C_{12} (e_1 + e_3), \tag{2.25b}$$
$$\sigma_3 = C_{11} e_3 + C_{12} (e_1 + e_2), \tag{2.25c}$$

and

$$\sigma_i = C_{44} e_i \quad i = 4, 5, 6. \tag{2.25d}$$

The energy of elastic deformation is then from Eq. (2.21) given by [6, 7, 28]

$$U = \frac{1}{2}C_{11}\left(e_1^2 + e_2^2 + e_3^2\right) + \frac{1}{2}C_{44}\left(e_4^2 + e_5^2 + e_6^2\right)$$
$$+ C_{12}\left(e_1 e_2 + e_1 e_3 + e_2 e_3\right) \tag{2.26}$$

and Eqs. (2.25) and (2.26) are related though Eq. (2.22).

It is found in the above examples that the highly symmetric system of the isotropic homogeneous medium is characterized by two independent elastic constants. For the cubic system the medium is described by three independent elastic constants. The parametrization of systems of other crystallographic symmetrizes are found in the literature [28].

Elastic Waves

The discussion of elastic waves is developed based on Eq. (2.16)

$$\rho\left(\vec{r}\right)\ddot{\vec{r}} = \rho\left(\vec{r}\right)\left[\ddot{s}_1\hat{x}_1 + \ddot{s}_2\hat{x}_2 + \ddot{s}_3\hat{x}_3\right]$$
$$= \left[\frac{\partial\sigma_{11}}{\partial x_1} + \frac{\partial\sigma_{21}}{\partial x_2} + \frac{\partial\sigma_{31}}{\partial x_3}\right]\hat{x}_1$$
$$+ \left[\frac{\partial\sigma_{12}}{\partial x_1} + \frac{\partial\sigma_{22}}{\partial x_2} + \frac{\partial\sigma_{32}}{\partial x_3}\right]\hat{x}_2 \tag{2.27}$$
$$+ \left[\frac{\partial\sigma_{13}}{\partial x_1} + \frac{\partial\sigma_{23}}{\partial x_2} + \frac{\partial\sigma_{33}}{\partial x_3}\right]\hat{x}_3$$
$$= \nabla \cdot \overset{\leftrightarrow}{\sigma}\left(\vec{r}\right).$$

Here Eq. (2.1b) relating the displacement of a point of the strained solid from its unstrain position is used. Applying Eq. (2.20) in Eq. (2.27) yields a wave equation for the modal elastic waves of the linear medium.

For example, in the case of the isotropic homogeneous medium described in Eq. (2.23) it follows that for the $\hat{x}_1$-components [6, 7, 28]

$$\rho\ddot{s}_1 = (2G + \lambda)\frac{\partial^2 s_1}{\partial x_1^2} + G\left(\frac{\partial^2 s_1}{\partial x_2^2} + \frac{\partial^2 s_1}{\partial x_3^2}\right) + (G + \lambda)\left(\frac{\partial^2 s_2}{\partial x_1 \partial x_2} + \frac{\partial^2 s_3}{\partial x_1 \partial x_3}\right) \tag{2.28}$$

with the equations for the $\hat{x}_2$- and $\hat{x}_3$-components given by the cyclic permutation of the subscripts $1, 2, 3$. These equations represent basic forms of the Helmholtz wave equation.

In a compact notation the equation of motion in Eq. (2.28) along with those for the other components are written as [7, 28]

$$\rho\ddot{\vec{s}} = (G + \lambda)\nabla\left(\nabla \cdot \vec{s}\right) + G\nabla^2\vec{s}$$
$$= (2G + \lambda)\nabla\left(\nabla \cdot \vec{s}\right) - G\nabla \times \left(\nabla \times \vec{s}\right). \tag{2.29}$$

Solutions of Eq. (2.29) can now be found in terms of longitudinal waves of the form $v = \nabla \cdot \vec{s}$ and transverse waves of the form $\vec{w} = \nabla \times \vec{s}$.

To this end, upon taking the divergence of Eq. (2.29) it is found that [6, 7, 28]

$$\rho \ddot{v} = (2G + \lambda) \nabla^2 v \tag{2.30}$$

defines the solutions of the longitudinal waves. Assuming $\vec{s}(\vec{r}) = \vec{s}_0 e^{i(\vec{k}\cdot\vec{r}-\omega t)}$ so that $v(\vec{r}) = v_0 e^{i(\vec{k}\cdot\vec{r}-\omega t)}$, it then follows from Eq. (2.29) that the longitudinal solutions have a dispersion relation of the form

$$\omega = \sqrt{\frac{2G + \lambda}{\rho}} k. \tag{2.31}$$

Similarly, taking the curl of Eq. (2.29) gives

$$\rho \ddot{\vec{w}} = G \nabla^2 \vec{w} \tag{2.32}$$

with a dispersion relation

$$\omega = \sqrt{\frac{G}{\rho}} k. \tag{2.33}$$

Assuming $\vec{s}(\vec{r}) = \vec{s}_0 e^{i(\vec{k}\cdot\vec{r}-\omega t)}$ then yields the transverse wave form $\vec{w}(\vec{r}) = \vec{w}_0 e^{i(\vec{k}\cdot\vec{r}-\omega t)}$.

In the case of the medium with cubic symmetry described in Eq. (2.25) it follows from Eq. (2.27) that for the $\hat{x}_1$-components [7, 28]

$$\rho \ddot{s}_1 = C_{11} \frac{\partial^2 s_1}{\partial x_1^2} + C_{44} \left(\frac{\partial^2 s_1}{\partial x_2^2} + \frac{\partial^2 s_1}{\partial x_3^2} \right) + (C_{12} + C_{44}) \left(\frac{\partial^2 s_2}{\partial x_1 \partial x_2} + \frac{\partial^2 s_3}{\partial x_1 \partial x_3} \right) \tag{2.34a}$$

with the equations for the $\hat{x}_2$- and $\hat{x}_3$-components given by

$$\rho \ddot{s}_2 = C_{11} \frac{\partial^2 s_2}{\partial x_2^2} + C_{44} \left(\frac{\partial^2 s_2}{\partial x_3^2} + \frac{\partial^2 s_2}{\partial x_1^2} \right) + (C_{12} + C_{44}) \left(\frac{\partial^2 s_3}{\partial x_2 \partial x_3} + \frac{\partial^2 s_1}{\partial x_2 \partial x_1} \right), \tag{2.34b}$$

and

$$\rho \ddot{s}_3 = C_{11} \frac{\partial^2 s_3}{\partial x_3^2} + C_{44} \left(\frac{\partial^2 s_3}{\partial x_1^2} + \frac{\partial^2 s_3}{\partial x_2^2} \right) + (C_{12} + C_{44}) \left(\frac{\partial^2 s_1}{\partial x_3 \partial x_1} + \frac{\partial^2 s_2}{\partial x_3 \partial x_2} \right), \tag{2.34c}$$

respectively. As with the isotropic homogeneous medium the solutions of Eq. (2.34) are subject to solution by the introduction of plane wave forms. The separation into longitudinal and transverse waves, however, is not possible except along certain symmetry axes of the cubic system.

As an example of Eq. (2.34), consider the propagation of elastic waves in the x_1-direction. Substituting the form $\vec{s} = \vec{s}_0 e^{i(kx_1-\omega t)}$ into Eq. (2.34) gives

$$\left(\omega^2 \rho - C_{11} k^2 \right) s_1 = 0 \tag{2.35a}$$

for the longitudinal waves, and

$$\left(\omega^2\rho - C_{44}k^2\right)s_2 = 0 \qquad (2.35b)$$

$$\left(\omega^2\rho - C_{44}k^2\right)s_3 = 0 \qquad (2.35c)$$

for the two transverse polarizations. For this high symmetry direction, the solutions are resolved into longitudinal and transverse polarizations.

2.1.2   THEORY OF FLUID MECHANICS

The treatment of fluids requires a modified theoretical formulation from that used in the study of solids. The new emphasis of the modifications arises from the ability of fluids to flow and from the different nature of the forces that act in the flowing medium. While solids experience small deformations and restoring forces, fluids do not. Fluids in fact may exhibit new features of long ranged translations and even turbulence. In addition, viscous interactions can become important elements to an accurate determination of the fluid flow. These considerations are now introduced in the reformulation of Newton's second law adapted to flow.

In the following, the motion of a fluid of density, $\rho(\vec{r}, t)$, is treated by studying the trajectory of an infinitesimal volume element of the fluid as it moves through space with a velocity $\vec{v}(\vec{r}, t)$. As the element moves in time through space its velocity changes as it experiences force interactions with the rest of the fluid and from external fields. During the motion the sum of the forces on the infinitesimal element of volume δV is given by $\vec{F} = \vec{f}(\vec{r}, t)\,\delta V$ where $\vec{f}(\vec{r}, t)$ is the force per volume at the volume element being studied.

Considering the motion of the infinitesimal element of fluid, δV, the velocity of the element changes according to [1, 2, 6, 7, 28]

$$\vec{v}\left(\vec{r} + \vec{v}dt, t + dt\right) = \vec{v}\left(\vec{r}, t\right) + \frac{\vec{f}\left(\vec{r}, t\right)dt}{\rho}, \qquad (2.36a)$$

where $\vec{r}$ is its position at time t as it moves to its new position $\vec{r} + \vec{v}(\vec{r}, t)\,dt$ at $t + dt$. Equation 2.36a is essentially the statement of Newton's Second Law of motion for an infinitesimal time displacement of the element. In addition, note that the velocity of the element on the left-hand side of the equation accounts for the change in its position in dt as it translates in space with velocity $\vec{v}(\vec{r}, t)$ as well as the change from dt arising directly form the action of the forces. Expanding the left side of Eq. (2.36a) in dt it follows that

$$\frac{\partial}{\partial t}\vec{v}\left(\vec{r}, t\right) + \left(\vec{v}\left(\vec{r}, t\right) \cdot \nabla\right)\vec{v}\left(\vec{r}, t\right) = \frac{\vec{f}\left(\vec{r}, t\right)}{\rho}, \qquad (2.36b)$$

where the left-hand side of Eq. (2.36b) is now the so-called convective derivative of $\vec{v}(\vec{r}, t)$.

The force per volume on the right-hand side of Eq. (2.36b) arises from a number of sources. One source is the pressure, $p(\vec{r})$, in the fluid. This is the force per area acting across an area in the fluid. In addition, there are viscous interactions of the fluid with itself. Other sources may come from the interactions with external gravitational and electromagnetic fields. In the presentation below, these latter external forces will not be treated.

First consider the self-interaction of the fluid arising from the pressure in the fluid. For a cubic infinitesimal volume element with edges dx_1, dx_2, dx_3, the change in the pressure force per area acting between the faces of the cube is related to the pressure differential by [1, 2, 6, 7, 28]

$$dp = \frac{\partial p}{\partial x_1} dx_1 + \frac{\partial p}{\partial x_2} dx_2 + \frac{\partial p}{\partial x_3} dx_3$$
$$= \nabla p \cdot d\vec{r} = -\vec{f} \cdot d\vec{r}. \tag{2.37}$$

Notice the minus sign on the far right arises as the force on a volume element is opposite the direction of the pressure increase. This projects the fluid element to regions of decreasing pressures. In addition, the pressure interaction in Eq. (2.37) represents an energy conserving force forming the focus of the Euler formulation of fluid mechanics. This is now discussed.

From Eqs. (2.36b) and (2.37) it then follows that, accounting for the pressure forces in a fluid [1, 2, 6, 7, 28],

$$\frac{\partial}{\partial t} \vec{v}(\vec{r}, t) + (\vec{v}(\vec{r}, t) \cdot \nabla) \vec{v}(\vec{r}, t) + \frac{\nabla p(\vec{r}, t)}{\rho} = 0. \tag{2.38}$$

This is one of the two equations needed to determine the flow of fluids in the limit of a conservation of energy.

Another important relationship regarding the fluid flow in the limit of an Euler fluid is obtained by considering the flow of mass through a fixed volume of space. In particular, the flow of mass into the volume minus the flow of mass out of the volume in a given time interval must equal the change of mass within the volume during this same time interval. As a consequence, mass is not created or destroyed within the volume. In this regard, consider the cubic volume element with edges dx_1, dx_2, dx_3, fixed in space. The change in the mass flowing through the cube is given by [6, 7, 28]

$$dm = \left(\frac{\partial \rho v_1}{\partial x_1} dx_1 dx_2 dx_3 + \frac{\partial \rho v_2}{\partial x_2} dx_2 dx_1 dx_3 + \frac{\partial \rho v_3}{\partial x_3} dx_3 dx_1 dx_2 \right) dt$$
$$= -\frac{\partial \rho}{\partial t} dx_1 dx_2 dx_3 dt, \tag{2.39}$$

where a net positive dm is seen to represent a net flow of matter out of the cubic infinitesimal volume element. Consequently, from Eq. (2.39) it follows that a statement of mass conservation is

$$\frac{\partial \rho}{\partial t} + \nabla \cdot (\rho \vec{v}) = 0. \tag{2.40}$$

Equation 2.40 along with Eq. (2.36b) or Eq. (2.38) then constitute the Euler formulation of fluid mechanics.

Combining Eqs. (2.36b), (2.38), and (2.40), it follows that the flow of momentum in the fluid can be represented by

$$\frac{\partial \rho \vec{v}}{\partial t} - \nabla \cdot \overset{\leftrightarrow}{\sigma} = 0. \tag{2.41a}$$

Here [7, 28]

$$\sigma_{ij} = -\rho v_i v_j - \delta_{ij} P, \tag{2.41b}$$

are the components of a symmetric second rank tensor. The flow is seen to be a conservative flow.

An improved characterization of the fluid involves the introduction of viscous forces. This results in energy dissipation and forms the basis of the Navier–Stokes formulation of fluid mechanics. A simple form for the viscous force as a perturbation on the system is supplied by the following form [6, 7, 28]

$$f_1 = \eta \frac{\partial v_1}{\partial x_2} \tag{2.42}$$

representing the force per area in the $\hat{x}_1$ direction for the plane perpendicular to $\hat{x}_2$ that creates the velocity gradient, $\frac{\partial v_1}{\partial x_2}$, in the fluid.

In the limit of an incompressible fluid (i.e., a fluid in which the density is constant), a force per area similar to that in Eq. (2.42) can be introduced into the dynamical equation in Eq. (2.41) by generalizing the stress tensor to the form [6, 7, 28]

$$\sigma_{ij} = -\rho v_i v_j - \delta_{ij} P + \eta \left(\frac{\partial v_i}{\partial x_j} + \frac{\partial v_j}{\partial x_i} \right). \tag{2.43}$$

From Eqs. (2.43) and (2.41a) the Navier–Stokes equation for an incompressible fluid is obtained as [6, 7, 28]

$$\rho \frac{\partial \vec{v}}{\partial t} + \rho \left(\vec{v} \cdot \nabla \right) \vec{v} = -\nabla P + \eta \nabla^2 \vec{v}. \tag{2.44}$$

In the limit that $\eta \to 0$, this reduces to the Euler equation of motion of a nonviscous fluid [7, 28]

$$\rho \frac{\partial \vec{v}}{\partial t} + \rho \left(\vec{v} \cdot \nabla \right) \vec{v} = -\nabla P \tag{2.45}$$

which is now valid for both compressible and incompressible fluids.

A simple example of a solution of the Navier–Stokes equation is the motion of a fluid between two infinite parallel plates. The bottom plate at $x_2 = 0$ is stationary and the top plate at $x_2 = h > 0$ moves in the $\hat{x}_1$-direction with a constant velocity u. For this system the solution for the steady state velocity between $0 \le x_2 \le h$ is given by

$$v_1 (x_2) = \frac{u x_2}{h} - \frac{h x_2}{2\eta} \frac{\partial P}{\partial x_1} \left(1 - \frac{x_2}{h} \right). \tag{2.46}$$

The speed of the fluid then becomes a function of the depth in the fluid between the two plates.

Two limits of interest for Eq. (2.46) are Couette flow for which $\frac{\partial P}{\partial x_1} = 0$ (i.e., there is no pressure change in the direction of the flow) so that

$$v_1(x_2) = \frac{ux_2}{h},$$ (2.47)

and Poiseuille flow for which $u = 0$ and $\frac{\partial P}{\partial x_1} < 0$ (i.e., there is a pressure decrease in the direction of the flow) so that [7]

$$v_1(x_2) = -\frac{hx_2}{2\eta}\frac{\partial P}{\partial x_1}\left(1 - \frac{x_2}{h}\right).$$ (2.48)

These solutions illustrate flow properties of the fluid system which are not found in solids.

Another important class of solution of Eqs. (2.44) and (2.45) involve wave motions of the fluid. Consider Eqs. (2.40) and (2.45) treated in the linear, small amplitude, limit of a compressible fluid in which [6, 7, 28]

$$\rho = \rho_0 + \rho_\epsilon,$$ (2.49)

where ρ_0 is the constant uniform homogeneous density of the unstressed system and ρ_ϵ is the density change due to wave perturbation of the system. In addition, the pressure changes in the system are assumed in the adiabatic limit to be given by the adiabatic pressure relationship [6, 7, 28]

$$P = K\rho^\gamma.$$ (2.50)

To first order in ρ_ϵ and the small velocity, $\vec{v}$, Eqs. (2.40) and (2.45) become [6, 7, 28]

$$\frac{\partial \rho_\epsilon}{\partial t} + \rho_0 \nabla \cdot \vec{v} = 0$$ (2.51a)

and

$$\rho_0 \frac{\partial \vec{v}}{\partial t} + s^2 \nabla \rho_\epsilon = 0$$ (2.51b)

in which $s^2 = \gamma \frac{P_0}{\rho_0}$ where P_o is the pressure of the unperturbed medium. It then follows from Eq. (2.51) that

$$\frac{\partial^2 \vec{v}}{\partial t^2} - s^2 \nabla (\nabla \cdot \vec{v}) = 0,$$ (2.52)

and assuming a solution of the form $\vec{v} = \vec{v}_0 e^{i(\vec{k}\cdot\vec{r} - \omega t)}$ yields the dispersion relation for longitudinal propagating waves in the fluid as

$$\omega^2 = s^2 k^2.$$ (2.53)

In addition, note that the curl of Eq. (2.52) leads to $\frac{\partial^2 \nabla \times \vec{v}}{\partial t^2} = 0$, which from our earlier discussions of solids, indicates that the system does not support transverse excitations. Consequently, only longitudinal waves exist in the fluid.

The above treatment can be generalized to include energy losses of longitudinal planewave modes due to dissipation in a compressible fluid. This involves keeping the $\eta \nabla^2 \vec{v}$ term [28] in Eq. (2.44). In this formulation, the $\eta \nabla^2 \vec{v}$ term now represents the losses in the system, and the wave solutions are obtained from the modified form of Eq. (2.51b) [6, 7, 28]

$$\rho_0 \frac{\partial \vec{v}}{\partial t} + s^2 \nabla \rho_\epsilon - \eta \nabla^2 \vec{v} = 0. \tag{2.54}$$

Proceeding as in Eqs. (2.51)–(2.53), it then follows that the longitudinal planewave modes are solutions of

$$\rho_0 \frac{\partial^2 \vec{v}}{\partial t^2} - \eta \nabla^2 \frac{\partial \vec{v}}{\partial t} - \rho_0 s^2 \nabla \left(\nabla \cdot \vec{v} \right) = 0. \tag{2.55}$$

This again represents the propagation of longitudinal waves with a dispersion relation of the form

$$\omega^2 \left[1 + i \frac{\eta k^2}{\omega \rho_0} \right] = s^2 k^2. \tag{2.56}$$

2.2   ELECTROMAGNETIC THEORY

In engineering applications of classical electrodynamics, the focus is on the properties of the two fields known as the electric field and the magnetic induction. These fields represent the electric and magnetic properties of a system composed of charges, currents, and material media. As such, the fields are ultimately defined by the four Maxwell equations which relate them to the electric charges and currents in the system as well as to the presence of any polarized or magnetized media. The Maxwell equations along with various constitutive equations (describing the interaction of the media with the fields) and the Lorentz force equations (describing the forces generated on particles and currents by the fields) constitute a completely defined problem in electrodynamics. Problems of this type are a common subject of technological importance [30].

An additional formulation developed in electrodynamics is that used to treat relativistic effects. In discussions of relativity it is often convenient and natural to combine the electric and magnetic fields into one electromagnetic field represented by a second rank tensor. This leads to an elegant formulation of electrodynamics based in two tensor equations which, nevertheless, are equivalent to the four Maxwell equations mentioned earlier. In addition, in a full relativistic theory care must be exercised in defining the charge and current densities and in handling the description of the media present. Although of less interest to technology, this formulation will also be summarized in the following discussions.

Relativistic considerations based on the second rank tensor formulation are not necessary for most practical applications of electrodynamics in technology. However, some recent work in technology has a focus in the design of materials which mimic the relativistic phenomena found in the full relativistic treatment as applied in special and general relativity. For this reason, some mention of the formulation will be made in the following development, with an end to the discussion of mimicking relativistic effects. In the following discussions, however, the focus will be mainly on practical engineering application of the four Maxwell equations which are developed in the engineering formulation of the theory.

In the following treatment, first a development of the theory of the fields, forces and energy in the engineering formulation is made. This is followed by a treatment of the Maxwell equations with some examples of solutions and a development of the constitutive relations necessary to describe polarized and magnetized media. Finally, a summary of the second rank tensor treatment of relativity is discussed.

2.2.1   FIELDS, FORCES, AND ENERGY

Two important fields in engineering applications are the electric field, $\vec{E}\,(\vec{r}, t)$, and the magnetic induction, $\vec{B}\,(\vec{r}, t)$. These arise from the free charges and currents of the system studied as well as from the modifying presence of various dielectric and magnetic media in the problem. In this formulation, the fields $\vec{E}\,(\vec{r}, t)$ and $\vec{B}\,(\vec{r}, t)$ then both enter into a complete description of the mechanics of the charges and currents interacting with them. The interaction between the particles and fields is completely specified by the Lorentz force which consequently represents the connection of electromagnetism with the rest of physical reality [30].

For the case of a particle of charge q and velocity $\vec{v}$ the Lorentz force takes the form [30]

$$\vec{F} = q\left(\vec{E} + \frac{\vec{v}}{c} \times \vec{B}\right) \tag{2.57a}$$

while in the case of a charge density, $\rho\,(\vec{r}, t)$, and current density, $\vec{J}\,(\vec{r}, t)$, the Lorentz force is given by

$$\vec{F} = \int d^3r \left(\rho\vec{E} + \frac{1}{c}\vec{J} \times \vec{B}\right). \tag{2.57b}$$

Both force expressions in Eq. (2.57) connect the fields with the dynamics of the particles in the system which in turn modifies the fields generated by the particle motion. As a consequence of these interactions, the fields and particles enter into an exchange of energies and momenta. During this process of exchange, energy and momentum are continually shuttled back and forth between the two systems of the interacting particles and the fields.

One of the important manifestations of the fields is their energy exchanges with the particles. These exchanges arise from the work done on the charges and currents by the fields as well as the work by the particles in their generation of the fields. In the processes of energy transfer the work done and received by the fields then represents the important exchange between the particles and fields. The work exchange in the system, however, can be shown to involve only one of the two electromagnetic fields. As seen in Eq. (2.57), the forces from the magnetic field are always perpendicular to the motion of the charge in the system. This arises from the cross product in the Lorentz force. Consequently, the magnetic induction does no work on the particles, and only the work done by the electric field enters into energy considerations.

In the energy exchange between particles and fields only the electric field is available to provide a work interaction between the fields and particles. This interaction arises in the form of a flow of energy in the system. In particular, the power flow in a system of particles mediated by the electric field is given in terms of the particle currents and electric interactions in the problem by [30]

$$\int d^3r \, \vec{J} \cdot \vec{E}.\tag{2.57c}$$

Notice that the energy flow represented in Eq. (2.57c) can easily be expressed in terms of a single particle current moving on a trajectory $\vec{r}(t)$ by $\vec{J}(\vec{r},t) = q\vec{v}\delta^3(\vec{r} - \vec{r}(t))$ or in the case of a continuum distribution of particles by a current density. Consequently, the focus in the following discussions will be on systems of many particles described by a continuous current distribution which is easily specified to single particle systems.

2.2.2   MAXWELL EQUATIONS

In the absence of dielectric and magnetic media the vacuum-air Maxwell equations are given by [30]

$$\nabla \cdot \vec{E} = 4\pi\rho,\tag{2.58a}$$

$$\nabla \times \vec{B} = \frac{4\pi}{c}\vec{J} + \frac{1}{c}\frac{\partial \vec{E}}{\partial t},\tag{2.58b}$$

$$\nabla \times \vec{E} = -\frac{1}{c}\frac{\partial \vec{B}}{\partial t},\tag{2.58c}$$

$$\nabla \cdot \vec{B} = 0,\tag{2.58d}$$

and relate the two fields only to the charges and currents. The equations portray the fields as being generated by one another and by the presence of net charges and currents within the problem. Along with the Lorentz force, Eq. (2.58) provide a complete description of the electrodynamics of the system of particles in free space.

Upon introducing electromagnetic media into the problem in Eq. (2.58), changes in the Maxwell equations are required. The changes arise from the charges and currents associated with the atoms and molecules composing the media which now act as new sources contributing their own fields to the problem. In general, the new fields from the media are easily accounted for in a statistical description of the electromagnetic properties of the media. This is done by introducing certain polarization and magnetization vector fields into the problem. With the introduction of the new vector fields, in addition to the Lorentz force, various constitutive relationships generated from materials science are needed. These relationships relate the polarization and magnetization fields to the totality of electromagnetic fields in the system.

To model the presence of polarizable dielectric and magnetic media requires the introduction of an electric polarization field, $\vec{P}(\vec{r}, t)$, and a magnetization field, $\vec{M}(\vec{r}, t)$. These vector fields, respectively, represent the electric dipole and magnetic dipole densities of the materials present in the problem and are used to obtain the fields contributed to the problem by these sources of fields. The fields from these materials are in addition to the fields from the charge and current densities which exist in the system independent of the presence of the polarizable and magnetic materials. The polarization and magnetization fields, however, depend on the total electromagnetic fields in the system arising from the charges, currents, and media. This dependence will be treated later, following a discussion of the Maxwell equations in the presence of media.

In particular, the introduction into the problem of the polarization and magnetization fields of such media yields Maxwell equations of the form [30]

$$\nabla \cdot \vec{E} = 4\pi \left(\rho - \nabla \cdot \vec{P} \right), \tag{2.59a}$$

$$\nabla \times \vec{B} = 4\pi \left(\frac{1}{c} \vec{J} + \nabla \times \vec{M} \right) + \frac{1}{c} \frac{\partial \left(\vec{E} + 4\pi \vec{P} \right)}{\partial t}, \tag{2.59b}$$

$$\nabla \times \vec{E} = -\frac{1}{c} \frac{\partial \vec{B}}{\partial t}, \tag{2.59c}$$

$$\nabla \cdot \vec{B} = 0. \tag{2.59d}$$

Here the polarization and magnetization vectors in Eqs. (2.59a) and (2.59b) enter the right-hand sides of the equations as additional sources of charge and current densities associated with the vectors $\vec{P}$ and $\vec{M}$. The term $-\nabla \cdot \vec{P}$ represents a polarization charge density arising from the spatial variation of the polarization while $c \nabla \times \vec{M}$ describes a magnetization current density associated with the spatial variation of the magnetization. In addition, the polarization vector is found to enter into the description of the displacement current on the far right-hand side of Eq. (2.59b).

A simplification of the Eq. (2.59) can be made with the introduction of some additional notation. Upon defining the auxiliary fields of the electric displacement, $\vec{D} \equiv \vec{E} + 4\pi \vec{P}$, and the magnetic field, $\vec{H} = \vec{B} - 4\pi \vec{M}$, the Maxwell equations in Eqs. (2.59a) and (2.59b) are rewritten as [30]

$$\nabla \cdot \vec{D} = 4\pi \rho, \tag{2.60a}$$

$$\nabla \times \vec{H} = \frac{4\pi}{c} \vec{J} + \frac{1}{c} \frac{\partial \vec{D}}{\partial t}, \tag{2.60b}$$

while Eqs. (2.59c) and (2.59d) remain unchanged. Now all of the four fields $\vec{E}$, $\vec{B}$, $\vec{D}$, $\vec{H}$ are related solely to the charge and current densities which are independent of the polarizable and magnetized media. This is a simplification because for many materials the new vector fields have simple relationships to the electric field and the magnetic induction. In addition, note that in the absence of electromagnetic media, $\vec{D} = \vec{E}$ and $\vec{H} = \vec{B}$ so that Eqs. (2.58) are recovered.

As an example, in many so-called linear media the vector fields $\vec{D}$, $\vec{P}$ and $\vec{H}$, $\vec{M}$ are related to the fields $\vec{E}$ and $\vec{B}$ by a set of linear relationships. These take a general form given by [30]

$$\vec{D} = (1 + 4\pi\chi_e)\,\vec{E} = \varepsilon\vec{E}, \tag{2.61a}$$

$$\vec{B} = (1 + 4\pi\chi_m)\,\vec{H} = \mu\vec{H}. \tag{2.61b}$$

Here the coefficients ϵ and μ are the permittivity and permeability, and the coefficients χ_e and χ_m are the electric and magnetic susceptibilities, respectively. They are determined by the properties of the polarizable and magnetic media where they can be generated as statistical averages of first principles treatments or they can be obtained experimentally as phenomenological parameters. Along with the constitutive relationships in Eqs. (2.61) and the appropriate material parameters, Eqs. (2.59) and (2.60) provide a complete description of the magnetic fields within the media of any particular electrodynamic system. Fortunately, many of the technological applications of electrodynamics involve linear media so that the discussions of nonlinear media will be reserved to be treated later in these presentations.

For a linear media the energy considerations in Eq. (2.57c) become particularly simple and can be expressed as a conservative flow of energy through the total system of fields, particles, and electromagnetic media. To this end, solving Eq. (2.60b) for the current density in terms of the fields and using this in Eq. (2.57c) it follows that power flowing between the components of the system can be rewritten into the form [30]

$$\int d^3r \vec{J} \cdot \vec{E} = \frac{1}{4\pi} \int d^3r \left[c\vec{E} \cdot \left(\nabla \times \vec{H} \right) - \vec{E} \cdot \frac{\partial \vec{D}}{\partial t} \right]$$

$$= -\int d^3r \left[\frac{\partial u}{\partial t} + \nabla \cdot \left(\frac{c}{4\pi} \vec{E} \times \vec{H} \right) \right]. \tag{2.62a}$$

In the far-right side of Eq. (2.62a) the energy density in the electric and magnetic fields given by [30]

$$u = \frac{1}{8\pi} \left(\vec{E} \cdot \vec{D} + \vec{B} \cdot \vec{H} \right) \tag{2.62b}$$

has been introduced in a term representing the change in electromagnet energy located within the fields of the system. Additionally, to achieve this separation in the second equality in Eq. (2.62a), Faraday's law and the vector identity $\nabla \cdot \left(\vec{E} \times \vec{H} \right) = \vec{H} \cdot \left(\nabla \times \vec{E} \right) - \vec{E} \cdot \left(\nabla \times \vec{H} \right)$ have been apply.

A convenient way of expressing Eq. (2.62) as a conservation law is to rewrite the integrand in the far right-hand side of Eq. (2.62a) as [30]

$$\frac{\partial u}{\partial t} + \nabla \cdot \vec{S} = -\vec{J} \cdot \vec{E}. \tag{2.63a}$$

Here the Poynting vector

$$\vec{S} = \frac{c}{4\pi}\left(\vec{E} \times \vec{H}\right) \tag{2.63b}$$

represents the flux of electromagnetic energy in space such that the power flow through an area $d\vec{A} = dA\hat{n}$, where $\hat{n}$ is a unit vector perpendicular to the plane of the area dA, is given by

$$dU = \vec{S} \cdot d\vec{A}. \tag{2.63c}$$

In addition, the right-hand side of Eq. (2.63a) specifies the work done on the fields by the charges and currents of the system. Consequently, Eq. (2.63a) is a statement of the conservation of energy as energy is transferred through space and exchanged between the charge and current sources in the system.

2.2.3   SOLUTIONS OF MAXWELL EQUATIONS: ELECTROMAGNETIC WAVES

A solution of the Maxwell equations for the propagation of electromagnetic waves in a region of permittivity, ε, and permeability, μ, in the absence of charges and currents can readily be obtained from Eqs. (2.59) and (2.60). Under these conditions the Maxwell equations take the form [30]

$$\nabla \cdot \vec{E} = 0, \tag{2.64a}$$

$$\nabla \times \vec{B} = \frac{\mu\varepsilon}{c}\frac{\partial \vec{E}}{\partial t}, \tag{2.64b}$$

$$\nabla \times \vec{E} = -\frac{1}{c}\frac{\partial \vec{B}}{\partial t}, \tag{2.64c}$$

$$\nabla \cdot \vec{B} = 0. \tag{2.64d}$$

Taking the curl of Eq. (2.64c), applying Eqs. (2.64a) and (2.64b), and the vector identity $\nabla \times \left(\nabla \times \vec{a}\right) = \nabla\left(\nabla \cdot \vec{a}\right) - \nabla^2\vec{a}$ yields the wave equation

$$\nabla^2\vec{E} - \frac{\mu\varepsilon}{c^2}\frac{\partial^2\vec{E}}{\partial t^2} = 0, \tag{2.65}$$

and taking the curl of Eq. (2.64b) yields a similar equation for $\vec{B}$.

Solutions of Eq. (2.65) are in the form of planewaves given by

$$\vec{E}\left(\vec{r}, t\right) = \vec{E}_0 e^{i\left(\vec{k}\cdot\vec{r} - \omega t\right)} \tag{2.66a}$$

$$\vec{B}\left(\vec{r}, t\right) = \vec{B}_0 e^{i\left(\vec{k}\cdot\vec{r} - \omega t\right)}, \tag{2.66b}$$

where Eqs. (2.64a) and (2.64d) require that $\vec{k} \cdot \vec{E}_0 = \vec{k} \cdot \vec{B}_0 = 0$. Upon substitution into Eq. (2.65) the dispersion relation for a material with $\mu\varepsilon > 0$ is given by

$$k^2 = \mu\varepsilon \frac{\omega^2}{c^2} \qquad (2.67)$$

and from Eq. (2.63b) the time-averaged Poynting vector is [30]

$$\vec{S} = \frac{c}{8\pi} \frac{\sqrt{\mu\varepsilon}}{\mu} \left| \vec{E}_0 \right|^2. \qquad (2.68)$$

2.2.4 RELATIVISTIC FORMULATION

The above provides a review of the standard engineering formulation of electrodynamics. This formulation can be easily rewritten in a relativistic format which is relevant to discussion of systems designed to mimic relativistic effects by means of technologically engineered materials. In the relativistic formulation the earlier theory is rewritten in terms of vectors and tensors which transform between space-time frames via the Lorentz transformation. For the case of special relativity, the Lorentz transformation is set to leave the proper time interval, denoted $d\tau$ and defined by

$$c^2 d\tau^2 = c^2 dt^2 - dx_1^2 - dx_2^2 - dx_3^2, \qquad (2.69)$$

invariant. Here the time and space coordinates on the right-hand side of Eq. (2.69) are in a particular space-time frame of the problem for space time points $x_\alpha = (ct, x_1, x_2, x_3)$.

In the relativistic formulation the electric and magnetic fields are incorporated in a second rank tensor representing the single electromagnetic field. The form of the electromagnetic field tensor can be written in terms of the earlier discussed fields as [30]

$$F^{\alpha\beta} = \begin{vmatrix} 0 & -E_x & -E_y & -E_z \\ E_x & 0 & -B_z & B_y \\ E_y & B_z & 0 & -B_x \\ E_z & -B_y & B_x & 0 \end{vmatrix}. \qquad (2.70)$$

This tensor formulation is significant as, under the transformations of special relativity, it allows the electric and magnetic induction components to mix together under a Lorentz transformation between coordinates moving relative to one another.

Equation (2.70) can now be used to rewrite the engineering theory in forms which simplify relativistic treatments. In terms of the electromagnetic field tensor the Maxwell equation in Eqs. (2.58a) and (2.58b) are written as [30]

$$\partial_\alpha F^{\alpha\beta} = \frac{4\pi}{c} J^\beta \qquad (2.71a)$$

and the homogeneous Maxwell equations in Eqs. (2.58c) and (2.58d) as [30]

$$\partial^\alpha F^{\beta\gamma} + \partial^\beta F^{\gamma\alpha} + \partial^\gamma F^{\alpha\beta} = 0. \tag{2.71b}$$

In these equations,

$$J^\alpha = \left(c\rho, J_x, J_y, J_z\right) \tag{2.72}$$

represents the charge-current 4-vector expressed in terms of charge densities and current densities in Eq. (2.58), and

$$\partial_\alpha = \left(\frac{1}{c}\frac{\partial}{\partial t}, \frac{\partial}{\partial x_1}, \frac{\partial}{\partial x_2}, \frac{\partial}{\partial x_3}\right) \tag{2.73}$$

represents the 4-gradient vector operator.

In addition to the field equations, the continuity relations, expressing the conservation of charges and current, takes the simple form [30]

$$\partial_\alpha J^\alpha = 0, \tag{2.74}$$

where the left-hand side of this equation represents the 4-divergence of the charge current four vector. Expressing the Lorentz force in Eq. (2.57) in the relativistic notations, it is also rewritten into the form [30]

$$\frac{dP^\alpha}{d\tau} = \frac{q}{c}F^{\alpha\beta}U_\beta \tag{2.75}$$

in terms of the 4-velocity $U_\alpha = \frac{dx_\alpha}{d\tau}$ and the 4-momentum $P^\alpha = mU^\alpha$ where $U^\alpha = \frac{dx^\alpha}{d\tau}$ and $x^\alpha = (ct, -x_1, -x_2, -x_3)$.

CHAPTER 3

Acoustic and Electromagnetic Scattering from Cylinders

In this chapter two basic scattering problems are considered which are of interest in the study of photonic and phononic crystals and metamaterials. Both problems involve the treatment of the scattering of waves from cylinder targets, one for acoustic waves in elastic media and the other for electromagnetic waves in dielectric media. The treatment of these simple systems provides a basis which is important in understanding the electromagnetic [30, 31] and acoustic scattering [28, 29] from features often encountered in nanophotonics systems. Additionally, they illustrate the variety of considerations that are needed in order to arrive at scattering solutions, respectively, from more general acoustic and optical targets. In this regard, basic scattering treatments such as those introduced here are often found to be useful steps in studies focused on the acoustics and electromagnetic properties of complex engineered materials. As such, they have been applied to understand the properties of the engineered media that are a focus of current technologies and which contribute an important topic of discussion later in the text.

Both cylindrical scattering problems are of two-dimensional systems, and as shall be seen later such problems have been extremely important in the development of the photonics and phononics of periodic engineered media and metamaterials. Two-dimensional problems have an advantage over problems in one- and three-dimensions of being complicated enough to give interesting results but not so complicated as to require a major effort in their resolution. In this regard, two-dimensional models have been useful in the development of waveguides and circuits for the channeling of acoustic or electromagnetic radiation. They have many important applications as well in the design of filters and in the development of sensors and antennas and in other basic optical and acoustic devices. In addition, they have found application in the study of various plasmonic surfaces and surface acoustic waves which are an interest of recent plasmonic and surface phononics technologies.

The problems that are the focus of this chapter are the scattering of acoustic [29] and electromagnetic radiation [30, 31] by cylinders targets composed of media with simple basic properties. These particular targets often occur as components in a variety of engineered materials employed in device technologies or at least offer approximations of the components in these applications. In the acoustic problem, a pressure wave in a fluid is scattered from a cylinder composed of an isotropic-homogeneous solid medium and the interest is in the determination of the scattered pressure wave in the fluid. In the electromagnetic problem, the scattering of an incident electromagnetic planewave from a dielectric cylinder is treated between two isotropic-homogeneous media. For both considerations the interest is in the determination of the spatial properties of the acoustic or electromagnetic wave scattered by the cylinder target.

Both of the studies are essentially boundary value problems which shall be seen to be somewhat similar in their determinations. They involve the treatment of different types of vector fields which nevertheless are both solutions of Helmholtz (wave) equations. The boundary conditions for the solutions in the acoustic and electromagnetic problems arise as statements regarding the equality or continuity of sets of components of these fields at the interface between the different media in the problem. In the process of evaluating the two problems, a comparison of the similarities and differences of the acoustic and electromagnetic treatments will ultimately be made. First the acoustic problem is solved, followed by the treatment of the electromagnetic problem, and these are concluded by some comparative remarks on the two studies.

3.1   ACOUSTICAL SCATTERING

For the following discussions, consideration is given of the scattering of acoustic waves propagating in a nonviscous fluid incident on an isotropic homogeneous solid cylinder [29]. (A schematic of the scattering geometry is presented in Fig. 3.1.) In the scattering process, an incident planewave propagates in the x_1–x_2 plane and is scattered by a solid cylinder of radius a aligned with its axis along the x_3-axis of coordinates. The incident planewave moves in the x_1-direction for a normal incidence on the cylinder axis and generates a scattered wave which radiates from the cylinder in directions normal to the cylinder axis. The propagation of the scattered wave, like that of the incident wave, is parallel to the x_1–x_2 plane with the details of its spatial distribution related to the properties of the scattering cylinder. In this regard, while the solid composing the cylinder can support both longitudinal and transverse waves, due to the acoustic nature of the fluid medium both the incident and scattered waves in the fluid are longitudinal excitations.

The problem outlined is essentially a boundary value problem involving the solutions of the acoustic wave equations in the cylindrical coordinates appropriate to the symmetry of the problem. It requires solving for the waves in the fluid and matching their solutions to those of the waves in the solid. In this process, a successful treatment requires the application to the various solutions in the different media of a combination of appropriate asymptotic scattering boundary conditions on the incident and scattered waves as well as matching conditions at the fluid-solid interface. In the end the outcome of the study is the generation of the radiated pressure fields of the scattered wave created by the interaction of the cylinder with the incident planewave field in the fluid [29].

In the following, some reviews are made first of the waves in the solid medium and then of the waves in the fluid. This is followed by a discussion of the boundary conditions and the full treatment of the scattering boundary value problem. A final comparison is made of the theory with some experimental results [29].

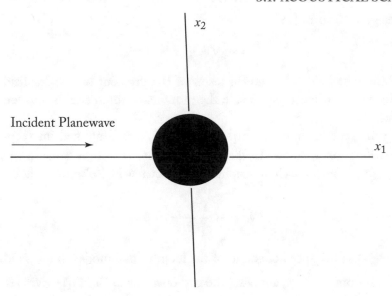

Figure 3.1: The scattering geometry for acoustic (electromagnetic) scattering from a cylinder of acoustic (dielectric) medium. The incident wave is propagating along the x_1-axis and is scattered propagating parallel to the x_1–x_2 plane.

3.1.1 ELASTIC WAVES IN THE ISOTROPIC HOMOGENEOUS SOLID

From the earlier discussions in Chapter 2, the equation for the motion of waves in the isotropic homogeneous solid is given as (see Eq. (2.29) in Chapter 2) [6, 7, 28, 29]

$$\rho\ddot{\vec{s}} = (2G + \lambda)\, \nabla\left(\nabla \cdot \vec{s}\right) - G\nabla \times \left(\nabla \times \vec{s}\right), \tag{3.1}$$

where $\vec{s}\left(\vec{r}\right)$ represents the displacement from equilibrium in the medium, ρ is the density of the medium, and $\{G, \lambda\}$ are Lame constants characterizing the elastic properties of the solid. The isotropic homogeneous medium represented in Eq. (3.1) is the most highly symmetric model of an elastic medium and ignores the symmetry of the crystal lattice which can affect the sound velocity as well as the polarization properties of the acoustic modes. As such, it tends to give an over simplified description of most solid media. Nevertheless, the model is found to be helpful as an illustration of the general semi-quantitative features of the acoustic scattering from a cylinder placed in a fluid medium, offering the simplest treatment of such a scattering problem.

Upon a consideration of the structure of the equation, it can be shown that the displacement vectors of the solid separate into propagating longitudinal and transverse waves. This separation arises from the isotropic homogenous nature of the medium, and a similar description in terms of longitudinal and transverse waves is not possible in a crystal medium with less symmetric elastic properties. A consequence of the separation is that the displacement fields can be expressed in the general form [29]

$$\vec{s} = -\nabla \psi + \nabla \times \vec{A}. \tag{3.2}$$

Here the displacement field is written in terms of the gradient of a scalar field and the curl of a vector field which, respectively, divide the motions of the medium into longitudinal and transverse waves.

Applying the representation in Eq. (3.2), Eq. (3.1) is simplified into separate standard Helmholtz equations for ψ and $\vec{A}$. In the case of ψ this separation is made by taking the divergence of Eq. (3.1). From such an application follows a Helmholtz equation for ψ of the form [29]

$$\nabla^2 \psi - \frac{1}{c_1^2} \frac{\partial^2 \psi}{\partial t^2} = 0, \tag{3.3}$$

where $c_1 = \sqrt{\frac{2G+\lambda}{\rho}}$ is the speed of sound of the longitudinal modes in the solid. Similarly, in the case of the components of $\vec{A}$, applying the curl operator to Eq. (3.1) gives [29]

$$\nabla^2 \vec{A} - \frac{1}{c_2^2} \frac{\partial^2 \vec{A}}{\partial t^2} = 0 \tag{3.4}$$

for the components of $\vec{A}$ representing the propagation of the transverse modes of the system. In Eq. (3.4) the speed of sound for the transverse modes is $c_2 = \sqrt{\frac{G}{\rho}}$ which from Eqs. (3.3) and (3.4) is seen to differ from that of the longitudinal modes.

The general solution for the waves in the solid is obtained from the solutions of Eqs. (3.3) and (3.4) as they are applied in Eq. (3.2) to generate the wave displacement fields in the solid. However, for a complete description of the dynamics of the solid, in addition to the displacement fields the strain and stress fields of the solid are needed. In this regard, the differential forms relating the displacement and strain fields were discussed in Chapter 2. The stress fields in the solid were also shown in Chapter 2 to be given by the linear stress-strain relations.

Following a development of the theory of elastic waves in the fluid a discussion of the Helmholtz solutions for both the solid waves and the fluid waves will be presented. These will then be used to develop the strain, stress, and pressure fields in these media.

3.1.2   ELASTIC WAVES IN THE FLUID

The acoustic waves in the nonviscous fluid surrounding the solid cylinder are also represented as solutions of a Helmholtz wave equation. For the case of a such a fluid medium only longitudinal waves are supported and are manifest as pressure waves which are incident on and scattered by the cylinder. These types of waves were preliminarily discussed in Chapter 2, and the focus of the following treatment will center on Eq. (2.45) from that chapter. Here, however, a different tack will be taken from the examples of fluid waves discussed in Chapter 2.

In the present case of the nonviscous fluid surrounding the solid cylinder, the Helmholtz equation of the waves propagating in the fluid is obtained starting from the Eq. (2.45) of Chapter 2. Considering small amplitude waves in which the nonlinear terms in Eq. (2.45) can be ignored, the equation relating the fluid velocity, $\vec{v}$, to the pressure gradient, ∇P, in a fluid of density, ρ_0, becomes [29]

$$\rho_0 \frac{\partial \vec{v}}{\partial t} = -\nabla P. \tag{3.5a}$$

Again, this equation is essentially Newton's Second Law, expressing the acceleration of a mass element of the fluid medium in terms of a pressure variation in the system. As with the earlier treatments of acoustic waves, in the following treatments the wave motion will arise from an application of boundary condition in the generation of the solutions of Eq. (3.5a).

If the displacement from equilibrium of the fluid due to a harmonic wave motion is described by the displacement vector, $\vec{s}$, then in terms of this vector Eq. (3.5a) becomes [29]

$$\rho_0 \frac{\partial^2 \vec{s}}{\partial t^2} = -\nabla P. \tag{3.5b}$$

From the form of Eq. (3.5b) it is seen that a displacement wave would be closely related to a pressure wave propagating through the system. In terms of the pressure representation, the pressure of the fluid is described by

$$P = P_0 + p, \tag{3.5c}$$

where P_0 is the pressure in the undisturbed fluid and p is the pressure change in the system caused by a wave passing through the fluid. In the following, a first focus will be on the development of a wave equation for the pressure waves in the fluid and this will ultimately be related to the displacement waves in the fluid.

To this end, substituting Eq. (3.5c) into Eq. (3.5b) and taking the divergence of the resulting equation yields

$$\rho_0 \frac{\partial^2 \nabla \cdot \vec{s}}{\partial t^2} = -\nabla^2 p \tag{3.6a}$$

which relates the Laplacian of the pressure variation to a time-dependent source term. In particular, notice that $\nabla \cdot \vec{s} = \frac{\partial s_1}{\partial x_1} + \frac{\partial s_2}{\partial x_2} + \frac{\partial s_3}{\partial x_3}$ is the dilatation or fractional change of a volume element in the fluid due to the displacement of the fluid. This dilatation is closely related to the pressure in the fluid and will be considered a small effect for the fluids being considered. Consequently, a reasonable assumption is that, for small pressure variations, the volume dilatation is linearly related to the pressure variation, i.e.,

$$\nabla \cdot \vec{s} = -\chi p. \tag{3.6b}$$

Here χ is a constant coefficient related to the bulk modulus of the fluid [28, 29].

When Eq. (3.6b) is applied to Eq. (3.6a), it follows that the pressure variations of the fluid are described by a Helmholtz equation of the form [28, 29]

$$\nabla^2 p - \frac{1}{c_3^2} \frac{\partial^2 p}{\partial t^2} = 0, \tag{3.7a}$$

where $c_3^2 = \frac{1}{\chi \rho_0}$ is the speed of acoustic waves in the fluid. In the case of a harmonic time dependence of the form $e^{-i\omega t}$, Eq. (3.7a) is seen to reduce to

$$\nabla^2 p + k_3^2 p = 0, \tag{3.7b}$$

where $k_3 = \frac{\omega}{c_3}$. Once the pressure has been set the displacement waves may, subsequently, be related to the pressure wave solutions of Eq. (3.7b) through Eq. (3.5). From an application of Eq. (3.5) this relationship is expressed by the form [29]

$$\vec{s} = \frac{1}{\rho_0 \omega^2} \nabla p. \tag{3.8}$$

Now that the Helmholtz equations for the waves in the fluid and in the solid have been obtained, it remains only to discuss the forms of their solutions and their assembly through the application of boundary conditions. First, a general discussion of the nature of the solutions of the Helmholtz equation will be given. Following this the solution will be formulated for the general scattering boundary value problem.

3.1.3   GENERAL SOLUTIONS OF THE HELMHOLTZ EQUATION FOR FLUIDS AND SOLIDS

All of Eqs. (3.3), (3.4), (3.7), and (3.8) are of the form of Helmholtz wave equations. Consequently, for a general study of the solutions of these equations in cylinder coordinates, the generic Helmholtz equation representing the wave equations is written as [29, 32]

$$\nabla^2 F - \frac{1}{c^2} \frac{\partial^2 F}{\partial t^2} = 0. \tag{3.9a}$$

In addition, for the case of a harmonic time dependence such that $F(r, \theta, z, t) = f(r, \theta, z) e^{-i\omega t}$, Eq. (3.9a) simplifies to the form [32]

$$\nabla^2 F + k^2 F = \frac{1}{r} \frac{\partial}{\partial r} \left(r \frac{\partial F}{\partial r} \right) + \frac{1}{r^2} \frac{\partial^2 F}{\partial \theta^2} + \frac{\partial^2 F}{\partial z^2} + k^2 F = 0, \tag{3.9b}$$

where $k^2 = \frac{\omega^2}{c^2}$. This later equation is important as an essential focus in determining the frequency dependence of the scattering from the cylinder.

In the following, Eq. (3.9) will be used for a general discussion of the relevant properties of the Helmholtz solutions which can then be specified to the forms of this equation in Eqs. (3.3), (3.4), (3.7), and (3.8). The properties reviewed will be those used to assemble appropriate solutions in the solid cylinder and the fluid which allow for matching the boundary conditions and ultimately determine the form of the complete scattering solution of the system.

Though the partial differential equation in Eq. (3.9) looks to be quite formidable, it can generally be simplified through the application of the method of separation of variables. The separation of variables method is based on assuming that the solutions of the partial differential equation can be expressed in terms of products of the solutions of ordinary differential equations. It is in fact based on reducing the partial differential equation into a set of ordinary differential equations which yield the functions composing the product solution. As an illustration, the method will now be applied to Eq. (3.9b).

In the simplified form of Eq. (3.9b), the study of the solutions of the partial differential equation are readily reduced to a treatment of three ordinary differential equations. This is accomplished by an application of the method of separation of variables in which $f(r, \theta, z)$ is represented as a product of three functions [32], i.e.,

$$f(r, \theta, z) = P(r) \Theta(\theta) Z(z). \tag{3.10}$$

By a substitution of this form into Eq. (3.9b), it can be shown that the solution of the partial differential equation is equivalent to the solutions of the set of ordinary differential equations [32]

$$\frac{d^2 Z}{dz^2} - l^2 Z = 0, \tag{3.11a}$$

$$\frac{d^2 \Theta}{d\theta^2} + m^2 \Theta = 0, \tag{3.11b}$$

$$r \frac{d}{dr} \left(r \frac{dP}{dr} \right) + \left[(k^2 + l^2) r^2 - m^2 \right] P = 0, \tag{3.11c}$$

where Θ is periodic in θ over intervals of 2π. As shall be seen later, the constants $\{l, m\}$, which are introduced during the process of converting the partial differential equation to a system of ordinary differential equations, occur as parameters in Eq. (3.10) and are used to match boundary conditions.

For problems involving normal incidence on the z-axis, a simplification occurs. In particular, the z dependence of the solutions drops out of the problem so that the relevant modes no longer depend on the z variable. Consequently,

$$f(r, \theta, z) = P(r) \Theta(\theta), \tag{3.12}$$

and in Eqs. (3.11a) and (3.11c) $l = 0$ so that Eq. (3.11c) becomes [32]

$$\rho \frac{d}{d\rho} \left(\rho \frac{dP}{d\rho} \right) + \left[\rho^2 - m^2 \right] P = 0, \tag{3.13}$$

where $\rho = kr$. The solution of the form in Eq. (3.12) will be the focus of the treatment which follows.

In the absence of a z-dependence, the solutions of Eq. (3.9b) for fix k^2 are composed from the set [32]

$$\{J_m(kr)\cos(m\theta),\quad J_m(kr)\sin(m\theta),\quad N_m(kr)\cos(m\theta),\quad N_m(kr)\sin(m\theta)\},\quad (3.14)$$

where $J_m(\rho)$ are Bessel functions, $N_m(\rho)$ are Neumann functions, and m ranges over the positive integers. Each member of the set in Eq. (3.14) constitutes a particular solution of Eq. (3.9b) and a form for the general solution of Eq. (3.9b) is then written as [32]

$$F(r,\theta,k) = \sum_{m=0}^{\infty} \{[a_m J_m(kr) + b_m N_m(kr)]\cos(m\theta)$$
$$+ [c_m J_m(kr) + d_m N_m(kr)]\sin(m\theta)\}, \quad (3.15)$$

where $\{a_m, b_m, c_m, d_m\}$ are set by boundary conditions. The remainder of the problem is to correctly match solutions which represent the scattering fields both outside and inside the cylinder.

3.1.4   BOUNDARY CONDITIONS

In regard to the boundary conditions, a number of conditions must be met inside the two media as well as at their interface. These conditions include: (1) in the fluid the solution must be represented as a combination of a planewave incident on the cylinder and an outgoing scattered wave radiating from the cylinder; (2) within the solid cylinder itself the wave solutions must be bounded and nonsingular; and (3) at the fluid-solid interface the displacements of both the solid and fluid materials normal to the interface must be equal, the pressure forces normal to the interface in both the fluid and solid must be equal, and the stresses in the solid parallel to the solid surface must be zero. These requirements account for all the conditions needed to uniquely define the coefficients of the solutions in Eq. (3.15).

For the scattering problem posed above, the incident planewave in the fluid is a pressure wave of the form [29]

$$p_{inc} = P_0 \exp(ik_3 x_1) = P_0 \exp(ik_3 r \cos\theta) = P_0 \sum_{m=0}^{\infty} \epsilon_m i^m J_m(k_3 r)\cos(m\theta). \quad (3.16)$$

Here, Eq. (3.16) is based on a standard expansion for a planewave propagating along the x_1-axis, $x_1 = r\cos(\theta)$, $\epsilon_0 = 1$, $\epsilon_{n>0} = 2$, and P_0 is the amplitude of the pressure wave in the fluid. It is seen in Eq. (3.16) that only $\cos(m\theta)$ terms enter into the expression for the incident wave. This is important as it sets the selection of the $\{\cos(m\theta),\ \sin(m\theta)\}$ terms composing the scattered waves in the fluid and of the waves in the solid.

Given the form in Eq. (3.16), the solution for the scattered waves radiating from the cylinder in the fluid are of the form [29]

$$p_{scat} = \sum_{m=0}^{\infty} c_m \left[J_m \left(k_3 r \right) + i N_m \left(k_3 r \right) \right] \cos \left(m\theta \right).$$ (3.17)

Here the combination of $J_m \left(k_3 r \right)$ and $N_m \left(k_3 r \right)$ in the brackets is chosen to give outgoing wave boundary conditions at $r \rightarrow \infty$ and c_m are determined by the boundary conditions at the fluid-solid interface. In addition, it should be noted that the $\cos \left(m\theta \right)$ terms are needed to match with the θ dependence of the planewave form in Eq. (3.16).

In addition to the pressure field the evaluation of the boundary conditions require the displacement fields of the fluid arising from the pressure wave. For the nonviscous fluid these are directly related to the pressure field of the fluid $p = p_{inc} + p_{scat}$ through Eq. (3.8). For matching the boundary conditions, of particular interest are the radial component of the displacement field given by

$$s_r^{fluid} = \frac{1}{\rho_0 \omega^2} \frac{\partial}{\partial r} p.$$ (3.18)

Applying Eq. (3.18) to the sum of Eqs. (3.16) and (3.17) gives the total radial displacement fields in the fluid as

$$s_r^{fluid} = \frac{P_0}{\rho_0 \omega^2} \sum_{m=0}^{\infty} \epsilon_m i^m \frac{d}{dr} J_m \left(k_3 r \right) \cos \left(m\theta \right)$$
$$+ \frac{1}{\rho_0 \omega^2} \sum_{m=0}^{\infty} c_m \frac{d}{dr} \left[J_m \left(k_3 r \right) + i N_m \left(k_3 r \right) \right] \cos \left(m\theta \right).$$ (3.19)

Here the first sum represents the contribution from the incident planewave and the second sum represents the contribution of the scattered wave generated by the interaction with the cylinder. These forms of the solutions in the fluid must now be matched to the solutions in the solid.

In the solid the forms of ψ and $\vec{A}$ needed to match up with the pressure waves in the fluid are [29]

$$\psi = \sum_{m=0}^{\infty} a_m J_m \left(k_1 r \right) \cos \left(m\theta \right)$$ (3.20a)

with $k_1 = \frac{\omega}{c_1}$ and

$$A_z = \sum_{m=0}^{\infty} b_m J_m \left(k_2 r \right) \sin \left(m\theta \right)$$ (3.20b)

with $k_2 = \frac{\omega}{c_2}$ where for the other cylinder components of $\vec{A}$, $A_r = A_\theta = 0$. These expressions for ψ and $\vec{A}$ do not involve the $N_m \left(k_1 r \right)$ and $N_m \left(k_2 r \right)$ terms as the Neumann functions are singular at $r = 0$ and there is nothing to account for such a singularity to exist in the solid medium of the cylinder. In addition, the θ dependence in Eq. (3.20) is chosen to match the correct boundary conditions with Eqs. (3.16) and (3.17). With the forms in Eqs. (3.16)–(3.20) the solutions are set to match the boundary conditions at the fluid-solid interface.

The displacement fields $\{s_r^{solid}, s_\theta^{solid}, s_z^{solid}\}$ in the solid are obtained from Eqs. (3.2) and (3.20) where the gradient and curl operators are expressed in cylinder coordinates. To this end, applying Eq. (3.20) in Eq. (3.2) it then follows that [29]

$$s_r^{solid} = \sum_{m=0}^{\infty} \left[\frac{mb_m}{r} J_m(k_2 r) - a_m \frac{d}{dr} J_m(k_1 r) \right] \cos(m\theta) \qquad (3.21a)$$

$$s_\theta^{solid} = \sum_{m=0}^{\infty} \left[\frac{ma_m}{r} J_m(k_1 r) - b_m \frac{d}{dr} J_m(k_2 r) \right] \sin(m\theta) \qquad (3.21b)$$

with

$$s_z^{solid} = 0. \qquad (3.21c)$$

For the matching of the boundary conditions, in addition to these displacement fields it is necessary to obtain the stress fields in the solid. These are obtained from Eq. (3.21) and the linear stress-strain relationships of the isotropic homogeneous solid. The linear stress-strain forms were discussed in Chapter 2 where they were expressed in cartesian coordinates. For the following applications they have been rewritten in cylinder coordinates.

3.1.5   APPLICATION OF THE BOUNDARY CONDITIONS AND SOLUTION OF THE PROBLEM

In the following, the three sets of boundary conditions are matched by the forms of solutions in Eq. (3.16)–(3.21). As a first boundary condition consider the conditions on the fluid and solid displacement fields normal to the fluid-solid interface. The condition that the normal components of displacement are equal at the fluid-solid interface involves the statement from Eqs. (3.19) and (3.21a) that

$$s_r^{fluid}(r = a) = s_r^{solid}(r = a). \qquad (3.22)$$

This follows from the requirement that the surface of the fluid does not separate from the surface of the solid under the passage of an acoustic wave between the two media. From Eq. (3.22) it follows that equating the right-hand sides of Eqs. (3.19) and (3.21a) and using the linear independence of the set $\{\cos(m\theta)\}$ yields the set of inhomogeneous equations [29]

$$
\begin{aligned}
k_1 a &J'_m(k_1 a) a_m - m J_m(k_2 a) b_m \\
&+ \frac{k_3 a}{\omega^2 \rho_0} \left[J'_m(k_3 a) + i N'_m(k_3 a) \right] c_m \\
&= -\frac{P_0 k_3 a}{\omega^2 \rho_0} \epsilon_m i^m J'_m(k_3 a)
\end{aligned}
\tag{3.23}
$$

for the coefficients $\{a_m, b_m, c_m\}$. Additional equations are needed for a complete determination of these coefficients, and these arise from the other two boundary conditions.

A second set of boundary condition equations is obtained from the equality at the interface of the pressure in the fluid to the normal component of stress in the solid. In the fluid the pressure wave in the fluid is obtained from Eqs. (3.16) and (3.17), and the stress in the solid is obtained from the linear stress-strain relations in Eq. (2.23) of Chapter 2 using the results for the strain from Eq. (3.21). In this way, written in terms of cylinder coordinates the pressure-stress boundary conditions take the form

$$
p_{inc}(r = a) + p_{scat}(r = a) = -\sigma_{rr}(r = a),
\tag{3.24a}
$$

where the linear stress-strain relations of the isotropic homogeneous solid in cylinder coordinates gives

$$
\sigma_{rr} = 2\rho c_2^2 \left[\frac{\sigma}{1 - 2\sigma} \nabla \cdot \vec{s} + \frac{\partial s_r}{\partial r} \right]
\tag{3.24b}
$$

for the Poisson's ratio, σ. Applying Eqs. (3.16), (3.17), and (3.21) in Eq. (3.24) yields another system of equations for the coefficients $\{a_m, b_m, c_m\}$. These are given by [29]

$$
\begin{aligned}
2\rho c_2^2 (k_1 a)^2 &\left[\frac{\sigma}{1 - 2\sigma} J_m(k_1 a) - J''_m(k_1 a) \right] a_m \\
&+ 2\rho c_2^2 m \left[k_2 a J'_m(k_2 a) - J_m(k_2 a) \right] b_m + a^2 \left[J_m(k_3 a) + i N_m(k_3 a) \right] c_m \\
&= -P_0 \epsilon_m i^m a^2 J_m(k_3 a).
\end{aligned}
\tag{3.25}
$$

A final system of equations for $\{a_m, b_m, c_m\}$ comes from the requirement that sheer stresses vanish at the fluid-solid interface. Specifically, this is stated in the appropriate form of the stress tensor in cylinder coordinates by

$$
\sigma_{r\theta}(r = a) = \sigma_{rz}(r = a) = 0
\tag{3.26}
$$

which upon substitution of Eq. (3.21) in the linear stress-strain relations of the solid yields a final set of linear equations for $\{a_m, b_m, c_m\}$. At the end of this process these equations are given by the system [29]

$$
\begin{aligned}
2m &\left[k_1 a J'_m(k_1 a) - J_m(k_1 a) \right] a_m \\
&= \left[m^2 J_m(k_2 a) - k_2 a J'_m(k_2 a) + (k_2 a)^2 J''_m(k_2 a) \right] b_m.
\end{aligned}
\tag{3.27}
$$

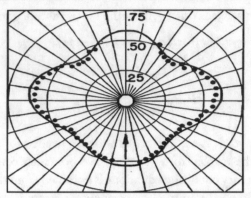

Figure 3.2: The acoustic scattering from a steel cylinder in water. Reprinted with permission from [Ref. [29]], Copyright 1951, Acoustic Society of America.

In this way the complete set of equations for $\{a_m, b_m, c_m\}$ are those of Eqs. (3.23), (3.25), and (3.27). This system of equations can be solved exactly in terms of closed form expressions, and the reader is referred to Ref. [29] for the detailed expressions of such a solution. The solution of these equations can also be obtained from a straightforward numerical solution of the system of equations. In either case the interesting result is to obtain the coefficients $\{c_m\}$ in terms of the parameters characterizing the elastic properties of the fluid and the solid and the parameters characterizing the incident pressure wave. Once the coefficients $\{c_m\}$ have been determined the scattered pressure wave obtained in the fluid is given directly from Eq. (3.17).

In this manner, from Eq. (3.17) it follows that for $r \to \infty$ the pressure of the scattered wave is given by [29]

$$|p_{scat}| \to \sqrt{\frac{2}{\pi k_3 r}} \left| \sum_{m=0}^{\infty} c_m e^{-im\frac{\pi}{2}} \cos(m\theta) \right|. \tag{3.28a}$$

Here the asymptotic form of the Hankel function of the first kind $H_m^{(1)}(s) = J_m(s) + i N_m(s) \to \sqrt{\frac{2}{\pi s}} e^{i(s - m\frac{\pi}{2} - \frac{\pi}{4})}$ as $s \to \infty$ has been applied in taking the limit in Eq. (3.28a).

3.1.6 AN EXAMPLE

An example of the scattering field in Eq. (3.28a) compared with experiment is given in Fig. 3.2. The figure shows the theory in Eq. (3.28a) compared to experimental data for the scattering of a steel cylinder in water. Specifically, the plot is made of [29]

$$\frac{1}{2} \sqrt{\frac{\pi k_3 r}{2}} \frac{|p_{scat}|}{P_0} \tag{3.28b}$$

vs. θ.

For these results the diameter of the steel cylinder is 0.032 inches for radiation at 1 mc/sec. In the computation the measured Young's modulus was 20×10^{11} dynes/cm^2, $k_3 a = 1.7$, $k_1 a = 0.45$, $\sigma = 0.28$, and $\rho = 7.7$ g/cm^3 [29]. The theory is seen to be well justified for this experimental system.

3.2   ELECTROMAGNETIC SCATTERING

The discussions in this section will treat the scattering of an electromagnetic planewave from a dielectric cylinder of circular cross section [30, 31]. (A schematic of the scattering geometry is presented in Fig. 3.1.) For simplicity both the medium of the cylinder and the medium in which it is embedded are lossless dielectrics that are homogeneous and isotropic. The incident planewave in the medium surrounding the cylinder propagates parallel to the x_1–x_2 plane and is scattered by the dielectric cylinder which is of radius a with an axis aligned along the x_3-axis of coordinates. The incident planewave moves in the x_1-direction normal incident on the cylinder axis, generating a scattered wave radiating away from the cylinder. The propagation of the scattered wave, like that of the incident wave, is parallel to the x_1–x_2 plane with the details of its spatial distribution related to the properties of the scattering cylinder. Unlike the acoustic wave scattering considered in the previous section, in both of the scattering media the electromagnetic waves are transverse waves. This has an effect on the boundary conditions needed at the interface between the two media.

In the scattering problem just outlined, the medium outside the cylinder is characterized by permittivity ε_1 and permeability μ_1, while the medium within the cylinder is characterized by permittivity ε_2 and permeability μ_2. Consequently, the Maxwell equations for the propagation of electromagnetic waves in the two regions of permittivities ε_1 and ε_2 and of permeabilities μ_1 and μ_2, respectively, are of the general form [30, 31]

$$\nabla \cdot \vec{E} = 0, \tag{3.29a}$$

$$\nabla \times \vec{B} = \frac{\mu_i \varepsilon_i}{c} \frac{\partial \vec{E}}{\partial t}, \tag{3.29b}$$

$$\nabla \times \vec{E} = -\frac{1}{c} \frac{\partial \vec{B}}{\partial t}, \tag{3.29c}$$

$$\nabla \cdot \vec{B} = 0, \tag{3.29d}$$

where $i = 1, 2$ for the media outside and insider the cylinder, respectively. From these sets of Maxwell equations follows the Helmholtz wave equations for transverse electric and magnetic waves. This transformation is done in the usual way applying curl operations between Eqs. (3.29b) and (3.29c) and expressing the results in terms of one of the two fields. In this manner, the wave equations for both fields are given by

$$\nabla^2 \vec{E} - \frac{\mu_i \varepsilon_i}{c^2} \frac{\partial^2 \vec{E}}{\partial t^2} = 0 \quad \text{for} \quad i = 1, 2, \tag{3.30a}$$

$$\nabla^2 \vec{B} - \frac{\mu_i \varepsilon_i}{c^2} \frac{\partial^2 \vec{B}}{\partial t^2} = 0 \quad \text{for} \quad i = 1, 2. \tag{3.30b}$$

For a homogeneous isotropic medium the solutions of Eq. (3.30) are in the form of transversely polarized planewaves expressed as

$$\vec{E}(\vec{r}, t) = \vec{E}_0 e^{i\left(\vec{k}_i \cdot \vec{r} - \omega t\right)} \quad \text{for} \quad i = 1, 2 \tag{3.31a}$$

$$\vec{B}(\vec{r}, t) = \vec{B}_0 e^{i\left(\vec{k}_i \cdot \vec{r} - \omega t\right)} \quad \text{for} \quad i = 1, 2, \tag{3.31b}$$

where from Eqs. (3.29a) and (3.29d) the transverse form of the waves follows from the conditions that $\vec{k}_i \cdot \vec{E}_0 = \vec{k}_i \cdot \vec{B}_0 = 0$. Substitution of Eq. (3.31) into Eq. (3.30), for $\mu_i \varepsilon_i > 0$ the dispersion relation in each medium is given by

$$k_i^2 = \mu_i \varepsilon_i \frac{\omega^2}{c^2} \quad \text{for} \quad i = 1, 2 \tag{3.32}$$

and from Eq. (2.68) the Poynting vector for the time averaged energy flux of the planewaves is

$$\vec{S} = \frac{c}{8\pi} \frac{\sqrt{\mu_i \varepsilon_i}}{\mu_i} \left| \vec{E}_0 \right|^2 \quad \text{for} \quad i = 1, 2. \tag{3.33}$$

The results in Eq. (3.31)–(3.33) are the most basic results in the electrodynamics of materials and in this sense precede those of the problem of the scattering from a single object. They, however, are a necessary starting point in the study of the cylinder scattering problem which shall be treated in the following discussions. The focus in the scattering problem will be the appropriate matching of the solutions within two different media which produce the scattered wave through an application of boundary conditions at the media interfaces.

In the following discussions, the scattering of planewaves in Eq. (3.31)–(3.33) by a dielectric cylinder are considered. After some preliminary discussions regarding the appropriate format for representing the incident planewaves, a separation of the scattering problem into a treatment of two different polarizations of the incident fields is made. In this regard, the scattering of incident waves with the electric field polarized parallel to the cylinder axis will first be considered. This is followed by a treatment of the scattering of incident waves with the magnetic field polarized parallel to the cylinder axis. In conclusion, the relation of these results to more general properties of photonic crystals and metamaterials is treated.

3.2.1   INCIDENT SOLUTIONS IN THE REGION OUTSIDE THE CYLINDER

First consider the cylinder coordinate representation of the incident wave solutions in the region outside the scattering cylinder. Such a representation is most important in evaluating the boundary value problem for the determination of the scattered waves. In these considerations the region outside the cylinder is an isotropic homogeneous dielectric medium characterized by the parameters μ_1, ε_1, and the total excitations in the dielectric consist of the incident planewave and the waves scattered that are heavily influenced by the cylindrical symmetry of the problem. As all the dielectrics within the problem are real, there is no energy loss in the system so that the energy in the incident and scattered waves are equal.

For the considerations of the cylindrical symmetry, the incident planewaves are best expressed using a standard representation in Bessel functions and cosine functions [30–32]. This is well known in the literature, and in this representation the electric field of the incident planewave is written as

$$\vec{E}_{inc,0}\left(\vec{r},t\right) = \vec{E}_{inc,0}\left(\vec{r}\right)e^{-i\omega t} \tag{3.34a}$$

where

$$\vec{E}_{inc,0}\left(\vec{r}\right) = \vec{E}_{inc,0}\exp\left(ik_1 x_1\right) = \vec{E}_{inc,0}\quad\exp\left(ik_1 r\cos\theta\right)$$

$$= \vec{E}_{inc,0}\sum_{m=0}^{\infty}\epsilon_m i^m J_m\left(k_1 r\right)\cos\left(m\theta\right). \tag{3.34b}$$

Here $\vec{E}_{inc,0}$ is the amplitude of the electric field in the incident wave, and the wavenumber in Eq. (3.34b) is defined in Eq. (3.32). The form in Eq. (3.34) is completely general and expresses all possible polarizations of the electric field of the incident electromagnetic waves.

In regard to the electric field, the scattering problem can be divided into the treatment of two different polarized solutions. In one polarization the electric field of the electromagnetic waves is polarized parallel to the axis of the scattering cylinder. In the other polarization the magnetic field of the electromagnetic waves is polarized parallel to the axis of the scattering cylinder. The general solution for the scattering of electromagnet waves by the cylinder is then formed as a linear combination of the two polarized solutions. In the following, first the polarization with the electric field parallel to the cylinder axis is consider. After this the treatment of the polarization with the magnetic field polarized parallel to the cylinder axis is given.

3.2.2   ELECTRIC FIELD POLARIZED PARALLEL TO THE CYLINDER AXIS

Due to the translational symmetry of the cylinder scattering problem posed above, one set of solutions of the problem exists with electric field components polarized parallel to the x_3-axis. In the following, the solution of the scattering for the case of this polarization is treated. For this set of solutions, the x_3-component of the incident electromagnetic planewaves takes the form [32]

$$E_{inc,0}\left(\vec{r}\right) = E_0 \exp\left(ik_1 x_1\right) = E_0 \quad \exp\left(ik_1 r \cos\theta\right)$$

$$= E_0 \sum_{m=0}^{\infty} \epsilon_m i^m J_m\left(k_1 r\right) \cos\left(m\theta\right). \tag{3.35}$$

Here Eq. (3.35) is based on a standard expansion for a planewave propagating along the x_1-axis, $x_1 = r \cos(\theta)$, $\epsilon_0 = 1$, $\epsilon_{m \neq 0} = 2$, and E_0 is the amplitude of the electric wave in the medium surrounding the scattering cylinder. In Eq. (3.35) only $\cos(m\theta)$ terms enter into the expression for the incident wave, and, as per the discussions of acoustic scattering in Section 3.1 of this chapter, this sets the $\{\cos(m\theta), \sin(m\theta)\}$ terms composing Helmholtz equation solutions of the scattered waves in the system.

Given the form in Eq. (3.35), the solution of the Helmholtz equation for the scattered waves radiating from the cylinder in the region outside the cylinder are of the form

$$E_{scat} = \sum_{m=0}^{\infty} c_m \left[J_m\left(k_1 r\right) + i N_m\left(k_1 r\right)\right] \cos\left(m\theta\right)$$

$$= \sum_{m=0}^{\infty} c_m H_m^{(1)}\left(k_1 r\right) \cos\left(m\theta\right), \tag{3.36}$$

where $H_m^{(1)}(x)$ is the Hankel function of the first kind. As in the discussion of the acoustic scattering, the combination of $J_m(k_1 r)$ and $N_m(k_1 r)$ in the brackets, making a Hankel function, is of the form giving outgoing wave boundary conditions at $r \to \infty$, and c_m are determined by the boundary conditions at the cylinder interface. In Eq. (3.36) only the $\cos(m\theta)$ terms are needed to match with the θ dependence of the planewave form in Eq. (3.35).

In addition to the electric fields the evaluation of the boundary conditions requires the magnetic induction fields of the components of electromagnetic waves in the system. These are determined directly form the Maxwell equations. Specifically, in terms of Eq. (3.29c) and the total electric field [30–32]

$$E_{outside} = E_{inc,0} + E_{scat}$$

$$= E_0 \sum_{m=0}^{\infty} \epsilon_m i^m J_m\left(k_1 r\right) \cos\left(m\theta\right) + \sum_{m=0}^{\infty} c_m H_m^{(1)}\left(k_1 r\right) \cos\left(m\theta\right), \tag{3.37}$$

the radial and angular expressions for the magnetic induction are generated by

$$B_r = -\frac{ic}{\omega} \frac{1}{r} \frac{\partial E_z}{\partial \theta} \tag{3.38a}$$

and

$$B_\theta = \frac{ic}{\omega} \frac{\partial E_z}{\partial r}. \tag{3.38b}$$

Here the general relations in cylinder coordinates between the z component of the electric field and the radial and angular components of the magnetic induction are through Eq. (3.29c), and the z-axis of the cylinder coordinates has been chosen to correspond with the x_3-axis of the Cartesian coordinates.

Applying Eq. (3.38) to the sum of Eqs. (3.35) and (3.36), the total angular component of the magnetic induction outside the cylinder is represented as

$$\begin{aligned}
B_{\theta,outside} = &\frac{ic}{\omega} \sum_{m=0}^{\infty} E_0 \epsilon_m i^m \frac{d}{dr} J_m (k_1 r) \cos (m\theta) \\
&+ \frac{ic}{\omega} \sum_{m=0}^{\infty} c_m \frac{d}{dr} H_m^{(1)} (k_1 r) \cos (m\theta).
\end{aligned} \tag{3.39}$$

Here the first sum on the right-hand side is the angular component of the magnetic induction of the incident wave, and the second sum on the right-hand side is the angular component of the magnetic induction of the scattered wave.

In Eqs. (3.36) and (3.38) the first sums represent the contribution from the incident planewave and the second sums represents the contribution of the scattered wave generated by the interaction of the incident waves with the cylinder. These solutions outside the cylinder must be matched to the solutions inside the cylinder. This is done by applying the match conditions that the components of the electric field and magnetic inductions are continuous at the cylinder interface.

The Maxwell wave solutions within the cylinder are now discussed. In the cylinder the solutions of the Helmholtz equations for the electric field and magnetic induction are, for the E_3 component of the electric field, written as [30–32]

$$E_{inside} = \sum_{m=0}^{\infty} a_m J_m (k_2 r) \cos (m\theta) \tag{3.40a}$$

and, by applying Eq. (3.38b) to Eq. (3.40), for the angular components of the magnetic induction given by

$$B_{inside,\theta} = \frac{ic}{\omega} \sum_{m=0}^{\infty} a_m \frac{d}{dr} J_m (k_2 r) \cos (m\theta). \tag{3.40b}$$

These expressions do not include $N_m (k_2 r)$ terms as the Neumann functions are singular at $r = 0$ and there is nothing to account for such a singularity to exist in the solid medium of the cylinder.

In addition, the θ dependence in Eq. (3.40) is chosen to match the correct boundary conditions with Eq. (3.36) and (3.40). With the forms in Eqs. (3.35)–(3.40) the solutions are set to match the boundary conditions at the cylinder interface.

Applying the electromagnetic boundary conditions requires the continuity of the tangential components of the electric field at the interface of the scattering cylinder. This leads to the set of equations in m given by

$$J_m(k_2 a)\, a_m - H_m^{(1)}(k_1 a)\, c_m = \epsilon_m i^m E_0 J_m(k_1 a). \tag{3.41a}$$

The equations in Eq. (3.41a) are not a complete set for the determination of the coefficients $\{a_m, c_m\}$, however, a consideration of the behavior of the magnetic fields can be used to completely set the values of these coefficients. In this regard, the additional boundary condition requiring the continuity of the tangential components of the magnetic field at $r = a$ implies $\frac{1}{\mu_1} B_{outside,\theta} = \frac{1}{\mu_2} B_{inside,\theta}$. Applying this condition on Eqs. (3.39) and (3.40b) at $r = a$ gives the system of equations in m

$$\frac{1}{\mu_2}\frac{d}{dr}J_m(k_2 r)\, a_m - \frac{1}{\mu_1}\frac{d}{dr}H_m^{(1)}(k_1 r)\, c_m = \frac{1}{\mu_1}\epsilon_m i^m E_0 \frac{d}{dr}J_m(k_1 r). \tag{3.41b}$$

Along with Eq. (3.41a) this set yields a determination of the entire set $\{a_m, c_m\}$.

Solving Eq. (3.41) for the amplitudes of the components of the scattered wave yields

$$c_m = \frac{J_m(k_2 r)\dfrac{1}{\mu_1}\dfrac{d}{dr}J_m(k_1 r) - J_m(k_1 r)\dfrac{1}{\mu_2}\dfrac{d}{dr}J_m(k_2 r)}{\dfrac{1}{\mu_2}H_m^{(1)}(k_1 r)\dfrac{d}{dr}J_m(k_2 r) - \dfrac{1}{\mu_1}J_m(k_2 r)\dfrac{d}{dr}H_m^{(1)}(k_1 r)}\epsilon_m i^m E_0 \tag{3.42}$$

where the right-hand side of Eq. (3.42) is evaluated at $r = a$. It then follows from Eq. (3.36) that in the far field limit

$$E_{scat} = \sum_{m=0}^{\infty} c_m H_m^{(1)}(k_1 r)\cos(m\theta) \to \sqrt{\frac{2}{\pi k_1 r}}\sum_{m=0}^{\infty} c_m e^{i\left(-m\frac{\pi}{2}-\frac{\pi}{4}\right)}e^{i k_1 r}\cos(m\theta). \tag{3.43}$$

Again, in this limit the other scattered fields in the problem are obtained from Eq. (3.42) and the Maxwell equations.

3.2.3   MAGNETIC FIELD POLARIZED PARALLEL TO THE CYLINDER AXIS

A second set of solutions for the cylinder scattering problem exists with the magnetic field polarized parallel to the x_3-axis. These solutions, together with the those discussed earlier for electric fields polarized parallel to the to the x_3-axis, form a complete set of modes in which any electromagnetic waves propagating in the system can be written. In the following a study of the scattering modes for the case of this second polarization is treated.

The x_3-component of the incident electromagnetic planewaves for the set of solutions with magnetic fields along the cylinder axis has the form [30–32]

$$H_{inc,0} = H_0 \exp(ik_1x_1) = H_0\exp(ik_1r\cos\theta) = H_0 \sum_{m=0}^{\infty} \epsilon_m i^m J_m(k_1r)\cos(m\theta). \quad (3.44)$$

It should be noted that this expression is very similar to that in Eq. (3.35) for the electric field of incident planewaves polarized parallel to the cylinder axis. As with Eq. (3.35), Eq. (3.44) is based on a standard expansion for a planewave propagating along the x_1-axis, $x_1 = r\cos(\theta)$, and H_0 is the amplitude of the magnetic wave. The form is readily seen to be a solution of the Helmholtz equation in the medium outside the cylinder, and as in the discussions in Section 3.2.2 only $\cos(m\theta)$ terms enter into Eq. (3.44). This dependence on $\cos(m\theta)$, similar to the solution with electric fields polarized along the cylinder axis, sets the $\{\cos(m\theta), \sin(m\theta)\}$ terms composing Helmholtz equation solutions of the scattered waves in the system.

Given the incident field in Eq. (3.44), the solution for the scattered waves radiating from the cylinder are represented by solutions of the Helmholtz equation given by [30–32]

$$H_{scat} = \sum_{m=0}^{\infty} c_m H_m^{(1)}(k_1r)\cos(m\theta), \quad (3.45)$$

where $H_m^{(1)}(x)$ is the Hankel function of the first kind. As in the discussion of the electric scattering fields in Eq. (3.36), the solution to the Helmholtz equation of the form of the Hankel function of the first kind is chosen to satisfy outgoing wave boundary conditions at $r \to \infty$, and c_m are determined by the boundary conditions at the cylinder interface. In Eq. (3.45) only the $\cos(m\theta)$ terms are needed to match with the θ dependence of the planewave form in Eq. (3.44).

In addition to the magnetic fields the evaluation of the boundary conditions requires the electric fields of electromagnetic waves. These are directly related to the magnetic field $H_{outside} = H_{inc,0} + H_{scat}$ through Eq. (3.29b) by the relations

$$E_r = \frac{ic}{\omega\epsilon_1}\frac{1}{r}\frac{\partial H_z}{\partial\theta} \quad (3.46a)$$

and

$$E_\theta = -\frac{ic}{\omega\varepsilon_1}\frac{\partial H_z}{\partial r}. \quad (3.46b)$$

Applying Eq. (3.46) to the sum of Eqs. (3.44) and (3.45) gives the total angular component of the electric field outside the cylinder as

$$E_{\theta,outside} = -\frac{ic}{\omega\varepsilon_1}H_0 \sum_{m=0}^{\infty} \epsilon_m i^m \frac{d}{dr}J_m(k_1r)\cos(m\theta)$$
$$-\frac{ic}{\omega\varepsilon_1}\sum_{m=0}^{\infty} c_m \frac{d}{dr}H_m^{(1)}(k_1r)\cos(m\theta). \quad (3.47)$$

In Eqs. (3.45) and (3.47) the first sums represent the contribution from the incident planewave and the second sums represents the contribution of the scattered wave generated by the interaction with the cylinder. To obtain a complete solution of the problem, the solutions outside the cylinder must be matched to the solutions inside the cylinder. These match conditions are that the components of the electric field and magnetic inductions are continuous at the cylinder interface. The matching up of the solutions in these two regions under theses continuity conditions is now discussed.

In the cylinder the solution for the H_3 component of the magnetic field is

$$H_{inside} = \sum_{m=0}^{\infty} a_m J_m(k_2 r) \cos(m\theta). \tag{3.48}$$

From Eq. (3.29b) for the region inside the cylinder the electric and magnetic fields are related through

$$E_r = \frac{ic}{\omega \epsilon_2} \frac{1}{r} \frac{\partial H_z}{\partial \theta} \tag{3.49a}$$

and

$$E_\theta = -\frac{ic}{\omega \epsilon_2} \frac{\partial H_z}{\partial r} \tag{3.49b}$$

so that

$$E_{inside,\theta} = -\frac{ic}{\omega \epsilon_2} \sum_{m=0}^{\infty} a_m \frac{d}{dr} J_m(k_2 r) \cos(m\theta). \tag{3.50}$$

As in Section 3.2.1, these expressions do not include $N_m(k_2 r)$ terms as the Neumann functions are singular at $r = 0$ and there is nothing to account for such a singularity to exist in the solid medium of the cylinder. In addition, the θ dependence in Eq. (3.48) is chosen to match the correct boundary conditions with Eqs. (3.47) and (3.48). With the forms in Eq. (3.44)–(3.50) the solutions are set to match the boundary conditions at the cylinder interface.

Applying the boundary condition requiring the continuity of the tangential components of the magnetic field at the interface of the scattering cylinder lead to the set of equations in m given by [30–32]

$$J_m(k_2 a) a_m - H_m^{(1)}(k_1 a) c_m = \epsilon_{mi}{}^m H_0 J_m(k_1 a). \tag{3.51a}$$

In addition, the boundary condition requiring the continuity of the tangential components of the electric field at $r = a$ implies $E_{outside,\theta} = E_{inside,\theta}$. Applying this condition on Eqs. (3.47) and (3.50) at $r = a$ gives the system of equations in m

$$\frac{1}{\varepsilon_2} \frac{d}{dr} J_m(k_2 r) a_m - \frac{1}{\varepsilon_1} \frac{d}{dr} H_m^{(1)}(k_1 r) c_m = \frac{1}{\varepsilon_1} \epsilon_{mi}{}^m H_0 \frac{d}{dr} J_m(k_1 r). \tag{3.51b}$$

Solving Eqs. (3.51) for the amplitudes of the components of the scattered wave yields

$$c_m = \frac{J_m(k_2 r)\frac{1}{\varepsilon_1}\frac{d}{dr}J_m(k_1 r) - J_m(k_1 r)\frac{1}{\varepsilon_2}\frac{d}{dr}J_m(k_2 r)}{\frac{1}{\varepsilon_2}H_m^{(1)}(k_1 r)\frac{d}{dr}J_m(k_2 r) - \frac{1}{\varepsilon_1}J_m(k_2 r)\frac{d}{dr}H_m^{(1)}(k_1 r)}\epsilon_m i^m H_0 \qquad (3.52)$$

where the right-hand side of Eq. (3.52) is evaluated at $r = a$.

From Eq. (3.45) in the far field limit

$$H_{scat} = \sum_{m=0}^{\infty} c_m H_m^{(1)}(k_1 r)\cos(m\theta) \rightarrow \sqrt{\frac{2}{\pi k_1 r}}\sum_{m=0}^{\infty} c_m e^{i\left(-m\frac{\pi}{2}-\frac{\pi}{4}\right)}e^{ik_1 r}\cos(m\theta) \qquad (3.53)$$

and the other fields in this limit are obtained applying the Maxwell equations to Eq. (3.53). An example of this results in Eqs. (3.43) and (3.53) is now evaluated for a particular example of cylinder scattering.

3.2.4 EXAMPLES OF ELECTROMAGNETIC SCATTERING FROM CYLINDERS

In the following an evaluation of Eqs. (3.43) and (3.53) is made in the so-called hard scattering limit. In this limit the cylinder is considered to exclude the fields of the incident and scattered waves from entering the region inside the cylinder. In addition, the examples only treat scattering in which the wavelength of the radiation is large compared to the cylinder radius for incident electromagnetic waves outside of the cylinder and at normal incidence to its axis. The limit of hard scattering involves considerations of both perfect conducting cylinders (i.e., infinite conductivity, σ) and superconducting cylinders.

For an additional simplification, the region outside the cylinder is taken to be vacuum and the cylinders are formed of either perfect conductors or superconductors considered for frequencies at low gap energies. Studies of these systems are now presented.

Perfect Conductors

In the limit of a perfect conducting cylinder, the incident and scattered electromagnetic waves originate from outside of the cylinder and are limited by the skin depth of the cylinder medium to the surface region of the cylinder. For the perfect conductor the skin depth of the radiation, $\delta = \sqrt{c^2/2\pi\omega\sigma}$, goes to zero as $\omega\sigma \to \infty$ for $\omega \to 0$ [1, 2, 30]. This excludes the entry of the time-dependent fields generated outside the cylinder into the cylinder interior. Note that the perfect conductor, unlike superconductors, does not exclude magnetic fields from its interior which are present prior to taking the $\sigma \to \infty$ limit. This latter phenomenon is the subject of the Meisner effect which is addressed later.

For the case of incident waves with the electric field polarized parallel to the axis of the perfect conductivity cylinder, $E_{outside}$ must be zero at the surface of the cylinder. Under this condition it follows from Eq. (3.37) that [30–32]

$$\epsilon_m i^m J_m(k_1 a) E_0 + c_m H_m^{(1)}(k_1 a) = 0. \tag{3.54}$$

In addition, the perfect conductivity requires that the component of magnetic induction perpendicular to the surface of the cylinder is zero at the surface. This latter requirement on the magnetic induction also results in the same conditions expressed in Eq. (3.54).

From Eq. (3.54) the coefficients of the scattered wave are given by

$$c_m = -\frac{\epsilon_m i^m J_m(k_1 a)}{H_m^{(1)}(k_1 a)} E_0 \tag{3.55a}$$

so that the scattered field is represented as

$$E_{scat}(\vec{r}) = \sum_{m=0}^{\infty} c_m H_m^{(1)}(k_1 r) \cos(m\theta). \tag{3.55b}$$

Equations (3.55) completely characterize the scattered electric fields in the regions outside the cylinder, and the corresponding components of the magnetic induction are obtained from the Maxwell equations.

A comparison with the acoustic scattering presented in Fig. 3.2 can be made by looking at the nature of the scattered fields in the far field. For this comparison, Eq. (3.55) must be expressed in the limit that $r \to \infty$. This is done using the asymptotic form of the Hankel function [32] of the first kind $H_m^{(1)}(s) = J_m(s) + i N_m(s) \to \sqrt{\frac{2}{\pi s}} e^{i\left(s - m\frac{\pi}{2} - \frac{\pi}{4}\right)}$ as $s \to \infty$. Applying this asymptotic form, Eq. (3.55b) then becomes

$$E_{scat}(\vec{r} \to \infty) = \sqrt{\frac{2}{\pi k_1 r}} e^{-i\left(\frac{\pi}{4}\right)} \sum_{m=0}^{\infty} c_m e^{-im\frac{\pi}{2}} e^{i(k_1 r)} \cos(m\theta). \tag{3.55c}$$

As in the case of the pressure waves considered earlier, it then follows that for $r \to \infty$ the amplitude of the scattered wave is given by

$$|E_{scat}(\vec{r} \to \infty)| \to \sqrt{\frac{2}{\pi k_1 r}} \left| \sum_{m=0}^{\infty} c_m e^{-im\frac{\pi}{2}} \cos(m\theta) \right|. \tag{3.56}$$

In Fig. 3.3 a plot is presented of $\sqrt{k_1 r} \frac{|E_{scat}|}{E_0}$ from Eq. (3.56) as a function of θ. The result is obtained considering a perfect conducting cylinder in air for the case in which $k_1 a = 0.40$.

Superconductors

A related model is that in which the perfect conducting cylinder is replaced by a Type I superconducting cylinder [1–3]. Now the electromagnetic field is excluded due to Meisner effect

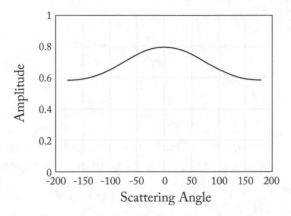

Figure 3.3: Plot of the Amplitude, $\sqrt{k_1 r}\,\frac{|E_{scat}|}{E_0}$, vs. the angle, θ, for the case in which $k_1 a = 0.40$.

considerations. These require that the magnetic induction be excluded from the bulk of the superconductor regardless of how the superconducting limit is approached. In the following some brief discussion is given of the electrodynamics of Type I superconductors after which these considerations are applied to aspects of the problem of cylinder scattering.

For the simple theoretical model presented in the following, the superconductor is treated in the London approximation. This involves considerations in which the London penetration length $\delta_L = \sqrt{mc^2/4\pi n e^2}$ is much larger that the coherence length, ξ. (This latter requirement is the standard characterizing property of Type I materials.) In addition, the cylinder medium will be taken to be well below the superconducting transition temperature so that $(T_c - T)/T_c \ll 1$, and the frequencies of radiation are considered to be within the superconducting gap so that $\hbar\omega \ll \Delta \ll k_B T_c$ where Δ is the gap order parameter [1, 2].

Under these restrictions the current density in the superconductor is given by [2, 3]

$$J\left(\vec{r}\right) = \left[-\frac{c}{4\pi\delta_L^2} + i\frac{\omega\sigma}{c} \right] \vec{A}\left(\vec{r}\right), \qquad (3.57)$$

where $\vec{A}\left(\vec{r}\right)$ is the vector potential and σ is the conductivity of the superconductor in its normal metal phase. (Note: For the limit that $\frac{\omega\sigma}{c} \to 0$, Eq. (3.57) reduces to the magnetostatic limit of the London equation.) In the following, Eq. (3.57) with $\mu = 1$ is the fundamental constitutive relation characterizing the currents and magnetic interactions in the superconductor. It will be used in the Maxwell equations to describe the electrodynamics of the superconducting medium.

In both the superconducting and vacuum media, the wave equations for the electric and magnetic fields are obtained in the usual way by taking the curls of the Amperes and Faraday laws and then using these same laws to obtain Helmholtz equations for each of the electric and magnetic induction fields. In this way it is found that [2]

$$\nabla^2 \vec{E} - \frac{1}{c^2} \frac{\partial^2}{\partial t^2} \vec{E} = \frac{4\pi}{c^2} \frac{\partial \vec{J}}{\partial t} \tag{3.58a}$$

and

$$\nabla^2 \vec{B} - \frac{1}{c^2} \frac{\partial^2}{\partial t^2} \vec{B} = -\frac{4\pi}{c} \nabla \times \vec{J}. \tag{3.58b}$$

Here in vacuum $\vec{J} = 0$ while Eq. (3.57) yields the current in the superconductor.

If the small loss terms in Eq. (3.57) are ignored, then in the superconductor [1–3]

$$\vec{J}(\vec{r}) \approx -\frac{c}{4\pi \delta_L^2} \vec{A}(\vec{r}). \tag{3.58c}$$

As a consequence, it follows from an application of the form for the current in Eq. (3.58c) and of the electrodynamics relation $\vec{E} = -\nabla \phi - \frac{1}{c} \frac{\partial \vec{A}}{\partial t}$ between the electric field and the scalar and vector potentials that [2]

$$\frac{\partial \vec{J}}{\partial t} \approx -\frac{c}{4\pi \delta_L^2} \frac{\partial \vec{A}}{\partial t} \approx \frac{c^2}{4\pi \delta_L^2} \vec{E}. \tag{3.59a}$$

Here the contribution for the scalar potential has been dropped so that $\frac{\partial \vec{A}}{\partial t} \approx -c\vec{E}$. In addition, from Eq. (3.58c) it follows that [1, 2]

$$\nabla \times \vec{J} = -\frac{c}{4\pi \delta_L^2} \nabla \times \vec{A}. \tag{3.59b}$$

The solutions to Eqs. (3.58) and (3.59) are obtained assuming a harmonic time dependence of the form $e^{-i\omega t}$ for the fields $\vec{E}$ and $\vec{A}$. From this, using the constitutive relations in Eq. (3.59) in Eq. (3.58), yields two Helmholtz equations for the electric and magnetic induction fields. These are given by

$$\nabla^2 \vec{E} - \frac{1}{c^2} \frac{\partial^2}{\partial t^2} \vec{E} = \frac{1}{\delta_L^2} \vec{E} \tag{3.60a}$$

and

$$\nabla^2 \vec{B} - \frac{1}{c^2} \frac{\partial^2}{\partial t^2} \vec{B} = \frac{1}{\delta_L^2} \vec{B}. \tag{3.60b}$$

Equations (3.60) can be expressed as Helmholtz equations written in terms of an effective dielectric constant, taking the forms [2, 3]

$$\nabla^2 \vec{E} + \varepsilon_s \frac{\omega^2}{c^2} \vec{E} = 0 \tag{3.61a}$$

$$\nabla^2 \vec{B} + \varepsilon_s \frac{\omega^2}{c^2} \vec{B} = 0 \tag{3.61b}$$

where $\varepsilon_s = 1 - \frac{c^2}{\delta_L^2 \omega^2}$ is the effective dielectric constant. It is seen that $\varepsilon_s \to -\infty$ as $\omega \to 0$ which effectively excludes both of the fields completely from the interior of the superconductor.

It is interesting to note that in the low frequency region for which $\omega \to 0$ Eqs. (3.61) take the form

$$\nabla^2 \vec{E} = \frac{1}{\delta_L^2} \vec{E} \tag{3.62a}$$

and

$$\nabla^2 \vec{B} = \frac{1}{\delta_L^2} \vec{B}. \tag{3.62b}$$

These are the London equations commonly used to describe the Meisner effect [1, 2].

The two equations in Eq. (3.61) represent the approach to the static field Meisner effect in the limit of low frequency radiation. The details of the scattering at these low frequencies are obtain by studying the scattered fields generated from an application of Eqs. (3.43) and (3.53). Later, the relevance of these results to the treatment of photonic crystals formed from super-conductors will be shown.

CHAPTER 4

Photonic and Phononic Crystals

In this chapter the basic theory of the properties of photonic and phononic crystals are discussed [4, 11, 20–24]. Photonic and phononic crystals are, respectively, periodic arrays of different dielectric or acoustic media. Whereas atomic and molecular crystals [1, 2, 4] are commonly thought of as periodic arrays of atoms and molecules, photonic and phononic crystals are arrays of macroscopic or mesoscopic bits of materials which are periodically ordered on a macro- or mesoscopic scale. The scales of the periodicity are then quite different between these various systems.

The periodicity of crystals has important consequences on the transport properties they exhibit [1, 2, 4, 11, 20–24]. For example, the periodicity of the positive ions in atomic and molecular crystals of metals, semiconductors, and insulators is responsible for the sequence of stop and pass bands in the electronic band structure of these materials. An importance of this is that the nature of the band structure is known to be a determining factor in setting the conductivity properties of the materials [1, 2, 4]. As shall be discussed later, similar stop and pass band effects, arising from periodicity, are found in the acoustic (optical) transport properties of phononic (photonic) crystals. In photonic and phononic crystals the size of the lattice and of the bits of material placed on the lattice can be tailored to the optical and acoustic radiations to be affected by their technology. In this sense photonic and phononic crystals offer a much greater flexibility in their applications than do molecular crystals [1, 2, 4, 11, 20–24]. They may be engineered structures whereas atomic crystals are generally much more difficult to tailor to specific applications.

The ordering in all the various crystal structures mentioned is made by repeatedly placing physical elements on the sites of a geometric lattice [1, 2, 4, 11, 20–22]. In the case of molecular crystals these elements are molecules which fashion their own lattice structures, while in photonics and phononics the features are, respectively, formed of dielectric or acoustic materials which are placed on an engineered lattice. For developing these periodic orderings, the geometric lattice of the crystals may be a one-dimensional, two-dimensional, or three-dimensional set of lattice points. In Fig. 4.1 for example are schematics representing a one-dimensional lattice of equal spaced points and an array of points of a two-dimensional square lattice. The photonics (acoustics) of a one-dimensional photonic (phononic) crystal is made from a periodic layering of slabs of dielectric (acoustic) materials on the one-dimensional lattice. Two-dimensional crystal systems may be formed, for example, as cylinders of dielectric or acoustic materials placed on the lattice sites of the square lattice. Similarly, the stacking of blocks of materials is a way of designing three-dimensional crystals.

Atomic crystals are interesting as their macroscopic properties are greatly affected by the periodic ordering of the atoms and molecules from which they are formed [1, 2, 4]. For example, the characteristic electronic and thermal electric properties of insulators, semiconductors, and metals are determined by the nature of the stop and pass bands opened in the electronic dispersion relations of each of these types of materials. These features, however, also influence the characteristic optical, magnetic, and thermal transport properties of the atomic crystals. As an added feature, the presence of impurities in otherwise crystalline systems cannot only modulate the properties set by the periodicity but can also introduce new physical properties in the materials that are not based on periodicity. In this regard, later it will be discussed how the flexibility in the manufacture of photonic and phononic crystals can be effective in making optical and acoustic waveguides and resonators. These features are introduced into the photonic or phononic crystals as sets of impurities.

Due to the periodicity of molecular, acoustic, and photonic crystals, the modal solutions of their excitations take specific forms which greatly affect the transport properties displayed by the system [1, 2, 4]. Specifically, the general form of the excitations propagating in these types of materials are composed of a planewave multiplying a function which is periodic in the lattice. In addition, the effects on the dispersion relations of this type of modal structure show up as a sequence of frequency stop and pass bands which are correlated with the wavevector of the planewave forms of the modal solutions for a sequence of different bands. A consequence of this is that excitations at frequencies in a stop band of a photonic or phononic crystal do not propagate through the bulk of the crystal, and propagation only occurs in the pass bands. As a result, excitations only move through the bulk of a photonic or phononic crystal at the sets of pass band frequencies and those at stop band frequencies quickly decay when passing into the bulk.

This pass-band stop-band structure of the bulk excitation spectrum of photonic and phononic crystals is at the root of many useful technological properties [4, 11, 20–22]. It allows for a variety of filtering, channeling, switching, and resonator applications, based on the ability to block certain frequencies from entering the bulk of the material.

To understand these properties, in the following a discussion of the general features of the band structure and the excitations of periodic systems is given. The discussions will first consider the effects of the translational symmetry of the periodic lattice on the properties of the systems. From these arise the spectrum of pass and stop bands in the frequency response of the systems. After this some considerations of group theory and the effects on the band structure of various point group symmetries arising in addition to the translational symmetries of the lattice are presented. The point group symmetries [32, 33] are shown to account for degeneracies in the dispersion relations of the systems at certain points of the dispersion relation of the excitations. This is followed by a discussion of technological developments based on the band structure properties of photonic and phononic crystals and the effects of impurities formed within them. As a conclusion, photonic crystal fibers and resonators are discussed followed by some indications of current applications of photonic and phononic crystals.

4.1   TREATMENT OF PERIODIC FUNCTIONS

In the following, discussions are made of the wavefunctions and dispersion relations in periodic media. First a detailed consideration is presented for one-dimensional waves. This is then followed by generalizations of these considerations to two- and three-dimensional systems. The treatment covers the properties of both acoustic and electromagnetic systems, focusing in both cases on the forms of the wavefunctions and the dispersion relations required by the symmetry properties of the media.

In both optical and acoustic systems, the modes of the system are described by eigenvalue problems based on the Helmholtz equation. Consequently, in the following discussions the focus is on the general Helmholtz eigenvalue problem and its symmetries.

4.1.1   ONE-DIMENSIONAL SYSTEMS

First consider the one-dimensional eigenvalue problem based on the Helmholtz equation [32] and defined over the one-dimensional periodic lattice in Fig. 4.1a. The general form of such a problem is written as

$$\frac{d^2}{dx^2} f(x) + \gamma w(x) f(x) = 0. \tag{4.1}$$

Here γ is an eigenvalue, and $w(x)$ is a periodic weight function such that

$$w(x - na) = w(x) \tag{4.2}$$

for n an integer and a the smallest repeat distance along the x-axis. In the crystal structures of interest to us, the equation is defined on the infinite x-axis and is subject to periodic boundary conditions at $x \to \pm\infty$.

Due to the periodic nature of $w(x)$, the solutions of Eq. (4.1) are restricted in form. To see this, the effects on Eq. (4.1) of a translation through a repeat distance which takes the periodic function $w(x)$ into itself is considered. The translation in the periodic space of $w(x)$ is made by the application of an operator T_{na} that shifts a general function $g(x)$ along the x-axis by an integer multiple of the smallest repeat distance, na, of the function $w(x)$. In this manner, $w(x)$ changes into $w(x - na)$ so that [1, 2, 4, 32, 33]

$$T_{na} w(x) = w(x - na) = w(x), \tag{4.3a}$$

and similarly, applying T_{na} to $f(x)$ in Eq. (4.1) yields

$$T_{na} f(x) = f(x - na). \tag{4.3b}$$

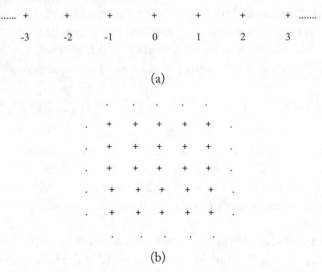

Figure 4.1: Infinite lattices: (a) one-dimensional periodic lattice with sites labeled sequentially by integers and (b) a two-dimensional square lattice. The lattice sites are indicated by + and the dots indicate a continuation to form a complete infinite lattice of points. The spacing between nearest neighbor points is a.

In order to find the effect of T_{na} on the total expression of Eq. (4.1), it only remains to see how the derivatives in Eq. (4.1) change under an application of T_{na}.

Now the effects of the translation operator on the derivatives in Eq. (4.1) are treated [32, 33]. Consider a translation by na applied to the derivative of the function $f(x)$. In this way, operating by T_{na} gives the transformation

$$T_{na}\frac{d}{dx}f(x) = \frac{d}{d(x-na)}f(x-na) = \frac{dx}{d(x-na)}\frac{d}{dx}f(x-na)$$

$$= \frac{1}{\dfrac{d(x-na)}{dx}}\frac{d}{dx}f(x-na) = \frac{d}{dx}f(x-na). \tag{4.4}$$

From a similar operation of T_{na} on the second derivative of $f(x)$, it then follows that

$$T_{na}\frac{d^2}{dx^2}f(x) = \frac{d^2}{dx^2}f(x-na). \tag{4.5}$$

In fact, the results in Eqs. (4.4) and (4.5) can be generalized to show that $T_{x_0}\frac{d^m}{dx^m}f(x) = \frac{d^m}{dx^m}f(x-x_0)$ for general integer m and real constant x_0. In this sense the derivative operators have a higher translational symmetry than the weight functions as they are invariant under all translations [32, 33].

Combining all of the above results, and applying (4.2)–(4.5) to the Helmholtz equation in Eq. (4.1), it follows that

$$T_{na}\left[\frac{d^2}{dx^2}f(x) + \gamma w(x) f(x)\right] = \frac{d^2}{dx^2}f(x-na) + \gamma w(x) f(x-na)$$
$$= \frac{d^2}{dx^2}T_{na}f(x) + \gamma w(x) T_{na}f(x) = 0. \tag{4.6}$$

Consequently,

$$\frac{d^2}{dx^2}T_{na}f(x) + \gamma w(x) T_{na}f(x) = 0, \tag{4.7}$$

and both $f(x)$ and $T_{na}f(x)$ are seen to be solutions of the same periodic Helmholtz equation. In addition, they are both solutions for the same value of the eigenvalue of the Helmholtz operator.

An understanding of the structure of the eigenfunctions and the dispersion relations of the modes of the wave equation follows from a consideration of Eqs. (4.1) and (4.7). Since both $f(x)$ and $T_{na}f(x)$ are eigenfunctions with the same eigenvalue, a convenient classification of the eigenfunctions of Eq. (4.1) is to relate them to the eigenfunctions of the translation operators, T_{na}. As shall now be seen this yields a statement about the general form for the wavefunctions of the eigenvalue problem in Eq. (4.1).

In this regard, consider now the eigenfunctions of Eq. (4.3). Specifically, begin by focusing on the solutions of

$$T_a f(x) = \sigma f(x), \tag{4.8}$$

where σ is the eigenvalue of the translation operator. The eigenfunctions of Eq. (4.8), as seen earlier, are also solutions of the periodic Helmholtz equation.

For n applications of T_a it follows that

$$T_{na} f(x) = \sigma^n f(x). \tag{4.9}$$

As a consequence, it is found from Eqs. (4.8) and (4.9) that σ must be of unit modulus, otherwise for $n \to \pm\infty$ the right-hand side of Eq. (4.9) would either collapse to zero or diverge, becoming infinite over the x-axis. In this manner it is seen that the only way the wavefunction can maintain its existence is if the normalized wavefunction maintains its normalization. Consequently, σ must be of the form

$$\sigma = e^{-ika}, \tag{4.10}$$

and for a one-dimensional lattice of $N \to \pm\infty$ sites the periodicity of the wavefunction in $2N$ requires that

$$k = \frac{n\pi}{Na} \quad \text{for} \quad n = 0, \pm1, \pm2, \pm3, \ldots. \tag{4.11}$$

From the discussions in Eqs. (4.7)–(4.11) it follows that, in order to be an eigenfunction both of Eqs. (4.1) and (4.9), $f(x)$ should have the form

$$f(x) = e^{ikx} u_k(x) \qquad (4.12)$$

where

$$T_{na} u_k(x) = u_k(x - na) = u_k(x) \qquad (4.13)$$

and the factor of σ in Eq. (4.10) arises from the planewave form

$$e^{ikx}. \qquad (4.14)$$

As a result of Eq. (4.13), it follows that $u_k(x)$ is a periodic function, and the general eigenfunctions of Eq. (4.1) are planewaves multiplying periodic functions. These planewave multipliers, in addition, are restricted by the conditions in Eq. (4.11).

Note that the eigenfunction in Eqs. (4.1) and (4.12) also exhibits a periodicity in $k-$ space. Specifically, under the transformation $k \rightarrow k - \frac{2\pi m}{a}$

$$e^{ikx} u_k(x) \rightarrow e^{i\left(k - \frac{2\pi m}{a}\right)x} u_{k - \frac{2\pi m}{a}}(x) = e^{ikx} v_k(x) \qquad (4.15)$$

where

$$v_k(x) = e^{-i\frac{2\pi m}{a}x} u_{k - \frac{2\pi m}{a}}(x) \qquad (4.16)$$

is a periodic function of x over the one-dimensional lattice [33]. Translated in this way the new wavefunction maintains the general form common to all of the solutions of the eigenvalue problem defined in Eq. (4.1). Consequently, the solution set of eigenvalues and their corresponding eigenfunctions are periodic in $k-$ space with the period having the smallest repeat distance, $\frac{2\pi}{a}$. It follows from these considerations that the eigenvalues, $\{\gamma_k\}$, corresponding to the eigenfunctions, $\{e^{ikx} u_k(x)\}$ of Eq. (4.15), obey the symmetry transformation [33]

$$\{\gamma_k\} = \left\{ \gamma_{k - \frac{2\pi m}{a}} \right\} \qquad (4.17)$$

under the transformation $k \rightarrow k - \frac{2\pi m}{a}$.

A consequence of the transformation properties in Eqs. (4.16) and (4.17) is that a complete set of solutions exist over any $k-$ space interval of width $\frac{2\pi}{a}$. The remaining solutions outside the interval are repetitions of those within the interval and may be obtained from the solutions within the interval by translations of $\frac{2\pi m}{a}$.

An Example

As an example of the wavefunction form in Eq. (4.12) and its eigenvalues in Eq. (4.17), the following presents a discussion of the system with $w(x) = 1$ followed by that of a system with a periodic $w(x)$. A comparison of the two solutions gives a useful contrast in understanding the nature and origins of the properties of periodic systems.

First note that if $w(x) = 1$ the weight function in Eq. (4.1) is unchanged by any translation vector along the x-axis. In this case, the wavefunctions of Eq. (4.1) are planewaves given by

$$f(x) = e^{ikx} \tag{4.18}$$

so that $u_k(x) = 1$ in Eq. (4.12) is also invariant under any translation vector along the x-axis. In addition, from Eq. (4.1) the eigenvalues corresponding to Eq. (4.18) are

$$\gamma_k = k^2. \tag{4.19}$$

It should be noted that the forms in Eqs. (4.18) and (4.19) are consistent with the earlier discussions of periodic systems because every shift along the x-axis is allowed for the $w(x) = 1$ system so that $a \rightarrow 0$ is the smallest transformation of the system in its lattice. Applied to the $k-$ space transformation, $k \rightarrow k - \frac{2\pi m}{a} \rightarrow k - \infty$ which leaves only Eq. (4.19) as the dispersion relation defined over the infinite region of $k-$ space.

Next consider the effects of introducing a periodic perturbation into the system. A simple example of a nonzero periodicity in the system of Eq. (4.1) is made by taking a weight function of the form

$$w(x) = \left[1 + \delta \cos\left(\frac{2\pi}{a}x\right)\right] \tag{4.20}$$

for $0 < \delta \ll 1$, and where a is the smallest repeat distance. For a consideration of Eq. (4.20) it is seen that in the $a, \delta \rightarrow 0$ limit the results in Eqs. (4.18) and (4.19) are regained while for $a \neq 0 \, \delta \ll 1$ the modifications of the periodic system are displayed for examination.

For the limit that $a \neq 0, \delta \rightarrow 0$ the eigenvalues of the solutions of Eq. (4.17) in the periodic system approach the limiting form

$$\gamma_k = \sum_{n=-\infty}^{\infty} \left(k - \frac{2\pi n}{a}\right)^2. \tag{4.21}$$

This is just the sum of a series of eigenvalues in Eq. (4.19) for the $w(x) = 1$, each shifted in $k-$ space by $\frac{2\pi n}{a}$. For γ_k in Eq. (4.21) it then follows that $\gamma_{k-\frac{2\pi m}{a}} = \gamma_k$, satisfying the condition in Eq. (4.17).

In Fig. 4.2 a plot of Eq. (4.21) containing the basic interval $-\frac{\pi}{a} \leq k \leq \frac{\pi}{a}$ is given, exhibiting a complete set of solutions for the $\delta \rightarrow 0$ limit of the system in Eqs. (4.1) and (4.20). This dispersion is now modified by the introductions of the $0 < \delta \ll 1$ term in Eq. (4.20).

Introducing the second term in Eq. (4.20) it is found that, to leading order in δ, the corrections to the eigenvalue spectrum in Eq. (4.21) open a series of stop bands in the dispersion relation of the $\delta \rightarrow 0$ system presented in Fig. 4.2. These stop bands are found to be gaps in the eigenvalue spectrum opened at the $k = -\frac{\pi}{a}$ and $k = \frac{\pi}{a}$ edges in Fig. 4.2. Excitations in the spectrum between these upper and lower edges of the band gaps do not propagate in the lattice.

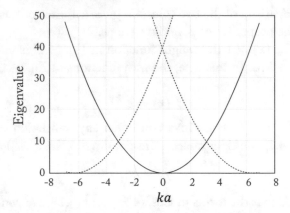

Figure 4.2: The normalized eigenvalue spectrum, $\gamma_k a^2$, from Eq. (4.21) plotted vs. ka.

As an example of the opening of one of the stop bands, consider the lowest solutions at the $k = \frac{\pi}{a}$ right-hand edge of Fig. 4.2. About this point, for k of the form $k = \frac{\pi}{a} + \Delta k$ in the $\frac{\Delta k}{\pi/a} \ll \delta \ll 1$ limit, the dispersion takes the form

$$\gamma_{\frac{\pi}{a}+\Delta k}^{(\pm)} \cong \left(\frac{\pi}{a}\right)^2 \left[1 \pm \frac{1}{2}\delta\right] + (\Delta k)^2 \left[1 \pm \frac{4}{\delta}\right]. \tag{4.22a}$$

It is seen from Eq. (4.22a) that at $\Delta k = 0$ the upper edge of the band gap is located at

$$\gamma_{\frac{\pi}{a}}^{(+)} \cong \left(\frac{\pi}{a}\right)^2 \left[1 + \frac{1}{2}\delta\right] \tag{4.22b}$$

and the lower edge of the band gap is located at

$$\gamma_{\frac{\pi}{a}}^{(-)} \cong \left(\frac{\pi}{a}\right)^2 \left[1 - \frac{1}{2}\delta\right]. \tag{4.22c}$$

Between these two band edges, in the region $\gamma_{\frac{\pi}{a}}^{(-)} < \gamma < \gamma_{\frac{\pi}{a}}^{(+)}$, there are no eigenvalue solutions. Away from the band edges in Eqs. (4.22b) and (4.22c), however, the modes at the upper band edge have increasing γ with $\Delta k \neq 0$ and the modes at the lower band edge have decreasing γ with $\Delta k \neq 0$. In this regard, as an additional final point, it is important to note that the dispersion relation is symmetric about $k = 0$ so that the same spectral gaps are opened at the left-hand side of the dispersion relation (i.e., at $k = -\frac{\pi}{a}$.) as are present on the right-hand (i.e., at $k = \frac{\pi}{a}$.) side.

On the whole, with regard to this new dispersion relation and its associated wavefunctions, the introduction of a nonzero δ perturbation in the system is found to fundamentally change the topology of the solution sets of Eq. (4.1). It shall be seen that the stop bands opened by the periodicity make a new variety of physical responses available to the system.

4.1.2 TWO-DIMENSIONAL AND THREE-DIMENSIONAL LATTICES

The above discussions generalize directly to periodic systems in higher dimensions. To see this, consider the eigenvalue problem

$$[H(\vec{r}) + \gamma w(\vec{r})] f(\vec{r}) = 0 \tag{4.23}$$

defined in either two or three dimensions. (See Fig. 4.1b for an example of the two-dimensional square lattice.) Here $H(\vec{r})$ is a differential form consisting solely of first and second derivatives, $w(\vec{r})$ is a weight function, and γ is the eigenvalue. The weight function in Eq. (4.23) is now taken to be periodic in either two- or three-spatial dimensions [32].

An example of such a problem in two dimensions is the case of the Helmholtz equation for electromagnetic waves. In terms of Eq, (4.23) this is given by

$$H(\vec{r}) = \nabla^2, \tag{4.24a}$$
$$w(\vec{r}) = \varepsilon(\vec{r}), \tag{4.24b}$$

and

$$\gamma = \frac{\omega^2}{c^2} \tag{4.24c}$$

where $\varepsilon(\vec{r})$ is a periodic dielectric function. Applied in Eq. (4.23) these operators define the problem generating the modes and dispersion relation in a photonic crystal. Similarly, a problem in two-dimensional acoustics involving the case of the Helmholtz equation for acoustic waves is defined by

$$H(\vec{r}) = \nabla^2, \tag{4.25a}$$
$$w(\vec{r}) = \frac{1}{v(\vec{r})^2}, \tag{4.25b}$$

and

$$\gamma = \omega^2 \tag{4.25c}$$

where $v(\vec{r})$ is the periodically varying speed of sound. Applied in Eq. (4.23) these operators define the problem for the modes and dispersion relation in a phononic crystal. Notice that in both of these systems, the three-dimensional acoustic or electromagnetic modes often require a full vector field to characterized them [11] so that

$$f(\vec{r}) \rightarrow \vec{f}(\vec{r}) \tag{4.26}$$

in Eq. (4.23). This feature will be discussed later.

For a periodic system in three dimensions there are three smallest linearly independent vectors $\{\vec{a}_1, \vec{a}_2, \vec{a}_3\}$ such that any spatial translation of the periodic function $w(\vec{r})$ into itself is made by the translation vector [1, 2, 4]

$$\vec{\vec{T}}_{nml} = n\vec{a}_1 + m\vec{a}_2 + l\vec{a}_3 \tag{4.27}$$

where n, m, l are integers. In fact, the translation vectors defined in Eq. (4.27) can be used to generate a point lattice with the same periodicity properties as $w(\vec{r})$. In this generation the set of translations in Eq. (4.27) are applied at any one specific point of the lattice to define the positions of the other points of the lattice. The lattice generated in this way is known as the direct lattice or Bravais lattice [1, 2, 4] and fills all of three-dimensional space with the array of lattice points. Similarly, in two dimensions there are two smallest linearly independent vectors $\{\vec{a}_1, \vec{a}_2\}$ such that the periodic function $w(\vec{r})$ translates into itself by the translation vector

$$\vec{\vec{T}}_{nm} = n\vec{a}_1 + m\vec{a}_2 \tag{4.28}$$

for integers n, m. In this way a plane of direct or Bravais lattice points is generated from the set of all translation vectors of the form in Eq. (4.28). In general, for thre dimensions there are found to be 14 different types of Bravais lattices while in two-dimensions there are 5 distinct types of Bravais lattice, and all of these are generated from the translation vectors in this manner.

In either of these two systems (i.e., three- or two- dimensional $w(\vec{r})$) the periodicity of the weight functions is characterized by the relationship

$$w\left(\vec{r} + \vec{\vec{T}}\right) = w(\vec{r}), \tag{4.29a}$$

where $\vec{\vec{T}} = \vec{\vec{T}}_\alpha$ for $\alpha = nml$ in three dimensions and $\alpha = nm$ in two dimensions. In addition, from the properties of derivatives discussed in Eq. (4.4), the operator containing the space derivatives must also transform as [1, 32, 33]

$$H\left(\vec{r} + \vec{\vec{T}}\right) = H(\vec{r}). \tag{4.29b}$$

The invariance of $H(\vec{r})$ in Eq. (4.29b) is a consequence of the invariance of derivatives under an arbitrary spatial translation and actually involves a greater degree of translation symmetry than that of periodicity. As a result of these two translation properties it follows that

$$H(\vec{r}) + \gamma w(\vec{r}) \tag{4.29c}$$

is periodic over the direct lattice, with the overall symmetry of Eq. (4.29c) being determined by that in Eq. (4.29a). The symmetry of Eq. (4.29c) in two and three dimensions, as in the one-dimensional case, acts to fix the general forms taken by the eigenfunctions and eigenvalues of problems in photonic and phononic crystals. This will now be addressed.

Under these conditions of translational symmetry in the direct lattice, it follows that if $f(\vec{r})$ is an eigenfunction of Eq. (4.23), then so is $\vec{T}_{nml}\, f(\vec{r}) = f\left(\vec{r} - \vec{\vec{T}}_{nml}\right)$ in three dimensions or $\vec{T}_{nm}\, f(\vec{r}) = f\left(\vec{r} - \vec{\vec{T}}_{nm}\right)$ in two dimensions. (Here we have applied the translation operators $\vec{T}_{nml}$ or $\vec{T}_{nm}$ to their respective three- and two-dimensional functions $f(\vec{r})$.) For example, in the two-dimensional system this is found from an application of $\vec{T}_{nm}$ to Eq. (4.23). In this way it follows that [33]

$$\vec{T}_{nm}\left[H(\vec{r}) + \gamma w(\vec{r})\right] f(\vec{r}) = \left[H(\vec{r}) + \gamma w(\vec{r})\right] \vec{T}_{nm} f(\vec{r}) = 0 \qquad (4.30)$$

with a similar sequence for a three-dimensional periodic system. By choosing the eigenvectors of Eq. (4.30) to also be eigenvectors of the translation operators a general form for the eigenvectors is obtained. This form is next addressed.

To arrive at the general form of the eigenvectors of the translation operators, consider the two-dimensional problem

$$\vec{T}_{nm} f(\vec{r}) = \sigma_{nm} f(\vec{r}) \qquad (4.31a)$$

where σ_{nm} is the eigenvalue, and notice that

$$\vec{T}_{nm} = \left(\vec{T}_{10}\right)^n \left(\vec{T}_{01}\right)^m \qquad (4.31b)$$

so that the eigenvalues of $\vec{T}_{nm}$ follow from those of $\vec{T}_{10}$ and $\vec{T}_{01}$. Based on a similar argument to that for the one-dimensional problem in Eqs. (4.9) through (4.11), the eigenvalues of $\vec{T}_{10}$ and $\vec{T}_{01}$ must have unit moduli. This follows as in the one-dimensional system the application of a translation operator to a periodic function an infinite number of times must not force it to either zero or infinity. As we shall see, the condition of unit modulus provides a great restriction on the general set of possible eigenvalues and eigenvectors.

A consequence of the unit moduli of the eigenvalues of the translation operators is that the basic unit translation operators $\vec{T}_{10}$ and $\vec{T}_{01}$ have eigenvalues $e^{-ik_1 a_1}$ and $e^{-ik_2 a_2}$, respectively, where $a_i = |\vec{a}_i|$ for $i = 1, 2$. From Eq. (4.31) it then follows that the eigenvalues of the general translation operators $\vec{T}_{nm} = \left(\vec{T}_{10}\right)^n \left(\vec{T}_{01}\right)^m$ are composed as a product of these and are represented as $\sigma_{nm} = e^{-i(nk_1 a_1 + mk_2 a_2)}$. The resulting structure of the eigenfunction of Eq. (4.23) needed in order to exhibit such properties under translation is that $f(\vec{r})$ should be of the form

$$f(\vec{r}) = e^{i\vec{k}\cdot\vec{r}} u_{\vec{k}}(\vec{r}) \qquad (4.32)$$

where

$$\vec{T}_{nm} u_{\vec{k}}(\vec{r}) = u_{\vec{k}}\left(\vec{r} - \vec{\vec{T}}_{nm}\right) = u_{\vec{k}}(\vec{r}). \qquad (4.33)$$

Consequently, $u_{\vec{k}}(\vec{r})$ is a periodic function, and the general eigenfunctions of Eq. (4.30) are planewaves multiplying periodic functions. The arguments given above generalize in a straight-forward manner to the three-dimensional case.

The functions in Eqs. (4.32) and (4.33) are set to be both eigenfunctions of Eq. (4.23) and of the translation vectors in Eq. (4.31). It only remains to find the nature of the periodic function in Eqs. (4.32) and (4.33) and the set of $\vec{k}$ over which these solutions are defined. Similar to the consideration in Eq. (4.11) for the one-dimensional problem, the set $\vec{k}$ is determined by applying periodic conditions to Eq. (4.32) at the infinite limits of the direct lattice. The detailed determination of the function $u_{\vec{k}}(\vec{r})$ itself requires a comprehensive numerical study which will be now discussed.

Periodic Part of the Wavefunction

An important component of the wavefunctions is the periodic function, $u_{\vec{k}}(\vec{r})$, arising from the periodicity of the weight function. To understand many of the properties of the system in Eq. (4.23) and its solutions requires an understand of the representation of these types of periodic functions as Fourier series. This has consequences for the understanding of the dispersive properties of the modes, their modal wavefunctions, and for the development of computational techniques for the generation of the solutions of Eq. (4.23). Arising from these considerations, a lattice in $k-$ spaced known as the reciprocal lattice is defined [1, 32, 33].

The periodic function $w(\vec{r})$ can be represented as a Fourier series of the standard from [1, 2, 4, 32, 33]

$$w(\vec{r}) = \sum_{\vec{G}} w_{\vec{G}} e^{i\vec{G}\cdot\vec{r}}. \tag{4.34}$$

This gives an expansion for $w(\vec{r})$ in terms of a sum over a set of planewaves defined over all space. In the process of generating Eq. (4.34), an important consideration is the set of planewaves used in the sum and the generation of the coefficients, $w_{\vec{G}}$. The particulars of these features show up in the physical properties of the system and are directly resolved from a consideration of the spatial translation properties of Eq. (4.34) over the direct lattice.

In this regard, consider the weight function under the application of a lattice translation operator, i.e.,

$$w\left(\vec{r} + \vec{\vec{T}}_{\alpha}\right) = \sum_{\vec{G}} w_{\vec{G}} e^{i\vec{G}\cdot\vec{r}} e^{i\vec{G}\cdot\vec{\vec{T}}_{\alpha}}. \tag{4.35}$$

For the equality of Eqs. (4.34) and (4.35) required by the translation symmetry of the system, it is necessary that $e^{i\vec{G}\cdot\vec{\vec{T}}_{\alpha}} = 1$ so that

$$\vec{G} \cdot \vec{\vec{T}}_{\alpha} = 2\pi p \tag{4.36}$$

for an integer p. Applying this condition to both the two- and three-dimensional systems, in the case of a three-dimensional system the solution set of vectors $\{\vec{G}\}$ satisfying Eq. (4.36) is given by $\vec{G}$ of the general form [1, 2, 4, 32, 33]

$$\vec{G}_{nml} = n\vec{b}_1 + m\vec{b}_2 + l\vec{b}_3, \tag{4.37}$$

while in two dimensions the solution set of vectors are of the general from

$$\vec{G}_{nm} = n\vec{b}_1 + m\vec{b}_2 \tag{4.38}$$

where in Eqs. (4.37) and (4.38) n, m, l are integers. In these expressions for $\vec{G}$ the vectors $\{\vec{b}_1, \vec{b}_2, \vec{b}_3\}$ in Eq. (4.37) are given by [1, 2, 4]

$$\vec{b}_1 = 2\pi \frac{\vec{a}_2 \times \vec{a}_3}{\vec{a}_1 \cdot \vec{a}_2 \times \vec{a}_3} \tag{4.39a}$$

$$\vec{b}_2 = 2\pi \frac{\vec{a}_3 \times \vec{a}_1}{\vec{a}_1 \cdot \vec{a}_2 \times \vec{a}_3} \tag{4.39b}$$

$$\vec{b}_3 = 2\pi \frac{\vec{a}_1 \times \vec{a}_2}{\vec{a}_1 \cdot \vec{a}_2 \times \vec{a}_3} \tag{4.39c}$$

and the vectors $\{\vec{b}_1, \vec{b}_2\}$ in Eq. (4.38) are given by

$$\vec{b}_1 = 2\pi \frac{\vec{a}_2 \times \hat{n}}{\vec{a}_1 \cdot \vec{a}_2 \times \hat{n}} \tag{4.40a}$$

$$\vec{b}_2 = 2\pi \frac{\hat{n} \times \vec{a}_1}{\vec{a}_1 \cdot \vec{a}_2 \times \hat{n}}, \tag{4.40b}$$

where $\hat{n}$ in Eq. (4.40) is a unit vector perpendicular to the plane containing the vectors $\{\vec{a}_1, \vec{a}_2\}$.

Note that the above considerations of the Fourier series are general to any periodic function in the direct lattice and that the vectors $\{\vec{b}_1, \vec{b}_2, \vec{b}_3\}$ form the set of shortest translations of an associated lattice in three-dimensional k−space into itself. This new lattice in k−space, defined in this way by $\{\vec{b}_1, \vec{b}_2, \vec{b}_3\}$, is termed the reciprocal lattice. Similarly, the set $\{\vec{b}_1, \vec{b}_2\}$ form the shortest translations of a reciprocal lattice in two-dimensional k−space and represents any periodic function defined over the periodic direct lattice. For both two and three dimensions the reciprocal lattices discussed here are in general different from their direct lattice counterparts.

The reciprocal lattice is seen to be a periodic array of points defined in k−space, and the translation vectors which translate the reciprocal lattice into itself are of the form [1, 2, 4]

$$\vec{G}_{nml} = n\vec{b}_1 + m\vec{b}_2 + l\vec{b}_3 \tag{4.41}$$

in three dimensions, and of the form

$$\vec{G}_{nm} = n\vec{b}_1 + m\vec{b}_2 \tag{4.42}$$

in two dimensions. Also related to these basis vectors of the reciprocal lattice translations is the smallest volume element with edges defined by $\left\{\vec{b}_1, \vec{b}_2, \vec{b}_3\right\}$ in the reciprocal lattice. This basic unit of volume is given by [1, 2, 4]

$$V_c = \left| \vec{b}_1 \cdot \vec{b}_2 \times \vec{b}_3 \right|, \tag{4.43}$$

and when translated in space by the complete set of vectors of the form in Eq. (4.36), the system of translated V_c fills all $k-$space. As shall be noted presently, each of the V_c in this partitioning of $k-$space contains a complete set of solutions for the eigenvalues and eigenvectors of the system. In two dimension a similar partition of the two-dimensional $k-$space is made by the area [1, 2, 4]

$$A_c = \left| \vec{b}_1 \times \vec{b}_2 \right|. \tag{4.44}$$

This area can be used to partition the two-dimensional $k-$space such that any single A_c can be shown to contain a complete set of solutions of the system. In this regard, V_c and A_c represent the modes of the three- and two dimensional solutions in the same manner as the interval $-\frac{\pi}{a} \leq k \leq \frac{\pi}{a}$ represents the modes of the one-dimensional problem.

An important point of symmetry is made by considering the effects of a shift in $\vec{k}-$space of the wavefunction forms in Eq. (4.32) through a reciprocal lattice translation vector as defined in Eqs. (4.42) and (4.41) for the two- or three-dimensional systems. Under this transformation

$$e^{i\left(\vec{k}+\vec{G}\right)\cdot\vec{r}} u_{\vec{k}+\vec{G}}\left(\vec{r}\right) = e^{i\vec{k}\cdot\vec{r}} v_{\vec{k}}\left(\vec{r}\right) \tag{4.45a}$$

where

$$v_{\vec{k}}\left(\vec{r}\right) = e^{i\vec{G}\cdot\vec{r}} u_{\vec{k}+\vec{G}}\left(\vec{r}\right) \tag{4.45b}$$

is periodic in the direct lattice. It is found that the translation in $\vec{k}-$space leaves the wavefunction in the form of a planewave times a periodic function so that the set of solutions has the same form before and after the translation. From this it follows that the eigenvector and eigenvalue sets of Eqs. (4.23) and (4.45) are periodic in the reciprocal lattice. Similar results are obtained for two-dimensional systems. Consequently, an elementary volume (defined in Eq. (4.43)) contains all of the three-dimensional solutions and an elementary area (defined in Eq. (4.44)) contains all of the two-dimensional solutions.

4.2   POINT GROUP SYMMETRIES

The translational symmetries are not the only types of symmetries operating in photonic and phononic crystals. In addition, a variety of so-called point group symmetries involving rotations, reflections, and space inversions can be involved in the invariant properties of the system. These symmetries do not include translations of the lattice but leave the crystal lattice untranslated under their application. For example, the points of the square lattice should be invariant under the set of rotations, reflections, etc. which leave a square invariant. Similarly, the symmetries of a cube should be found in a cubic lattice, the symmetries of a triangle in a triangle lattice, etc. In total, it has been shown that there are 32 different point groups consistent with the translational symmetries of a three-dimensional crystal. The total space group of a crystal is then composed of the set of translational symmetries, the set of point group symmetries, and the set of combined symmetries including both translations and point group elements. In total, there are 230 different space groups characterizing the symmetries of possible three-dimensional crystal systems. All of these various sets of symmetries contribute a great deal to the determination of the physical properties displayed by photonic and phononic systems [1, 2, 4].

While translational symmetries determine the general stop and pass band features found in modal dispersion relations of physical systems defined on a periodic lattice, the point group symmetries are very important in determining the degeneracies of the modal eigenvalues of the dispersion relations in $\vec{k}$−space. Furthermore, the coexistence of point group and translational symmetries puts a limitation on the sets of point group operations that can be found in crystal lattices. A consequence of this is that there are only 5 different general lattice types in two dimensions while in three dimensions there are only 32 different general lattice types. Each different lattice type in these classifications is composed as a unique set of point group and translation symmetries and each distinct collection of symmetries defines what is known as a Bravais lattice of the system.

4.2.1   ROTATIONAL SYMMETRY

In order to understand how translational symmetry can limit the set of point group symmetries, consider the set of rotational symmetries. In regard to rotations about crystal axes, it can be shown that only 1-fold, 2-fold, 3-fold, 4-fold, and 6-fold axes of rotational symmetry can be present in a crystal lattice. This acts as a major restriction on the types of structure available to crystalline medium. A simple proof of the limitations of rotational symmetry is now given [1, 2, 33].

Consider the system of points $\{A, B, A', A'\}$ in Fig. 4.3 which are contained within a plane of a periodic lattice. In the figure points A and B are two nearest neighbor points of the lattice that are separated by the distance r. Next assume that a set of axes of rotational symmetry exists perpendicular to the page and passing through each of the lattice points in the plane of the page.

Let the lattice of points in the page be invariant under rotations of $\pm\theta$ about the axes perpendicular to the page, i.e., the lattice is rotated into itself under such rotations. By rotating by θ about the axis passing through the point A, the point B moves to the new point B' which is separated from A by the distance r. Similarly, by rotating by $-\theta$ about the axis through the point B, the point A moves to the new point A' which is separated from B by the distance r.

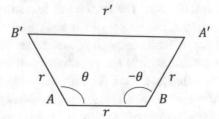

Figure 4.3: Four lattice points A, B, A', B' such that $\overline{AB}$ and $\overline{A'B'}$ are parallel. The points A' and B' are obtained by rotation by $-\theta$ and θ, respectively, as shown.

The set of points $\{A, B, A', B'\}$ form a trapezoid of lattice points having three sides of length r and a fourth side of length

$$r' = r - 2r\cos\theta = mr \tag{4.46a}$$

where m is an integer. This follows from the translational properties of the lattice in the horizontal direction. Consequently, the interval $\overline{A'B'}$ must have a length which is an integer multiple of the nearest neighbor distance r of the interval $\overline{AB}$ separating the points $\{A, B\}$.

It then follows from Eq. (4.46a) that

$$\cos\theta = \frac{1 - m}{2}, \tag{4.46b}$$

where $|\cos\theta| \leq 1$ for $m = 0, \pm 1, \pm 2, \ldots$ restricts the solutions of Eq. (4.46b) for θ to $\theta = \frac{2\pi}{n}$ with $n = 1, 2, 3, 4, 6$. These rotations are the signatures of the 1-fold, 2-fold, 3-fold, 4-fold, and 6-fold axes of rotational symmetry.

4.2.2   EFFECTS OF SYMMETRY OF THE EQUATIONS OF MOTION

Just as the eigenvalues and eigenfunctions of the wave equations can be chosen to be eigenvalues and eigenfunctions of the translational symmetries of the of the wave equations, they can also be chosen as simultaneously states of point group symmetries of the wave equations. In the case of the translational symmetries this simultaneity was demonstrated for one-dimensional problems in Eqs. (4.1)–(4.7) and in higher dimensions in Eqs. (4.23)–(4.30). There it was shown that in one-dimensions both the eigenfunction $f(x)$ and its translated form $T_{na}f(x)$ would be eigenfunctions corresponding to the same eigenvalues. A similar set of relations were also shown for the three- and two-dimensional systems involving an eigenfunction and its translated forms.

Now consider a set of lattice point group symmetries $\{O_\alpha\}$. These are sets of operations which under their application leave the operator of the eigenvalue problem unchanged.

For the case of the one-dimensional problem in Eq. (4.1) this implies that [1, 32, 33]

$$O_\alpha \left[\frac{d^2}{dx^2} f(x) + \gamma w(x) f(x) \right] = \frac{d^2}{dx^2} O_\alpha f(x) + \gamma w(x) O_\alpha f(x) = 0. \tag{4.47}$$

As in the case of translational invariance the $\{O_\alpha\}$ invariance relationships in Eq. (4.47) demonstrate that if $f(x)$ is an eigenfunction of the wave equation operator, then so is $O_\alpha f(x)$. They are both solutions of the same eigenvalue problem.

Similarly, applied to the higher dimensional systems, it follows from Eq. (4.23) that for a set of higher dimensional point group symmetries denoted by $\{O_\alpha\}$

$$O_\alpha \left[H(\vec{r}) + \gamma w(\vec{r})\right] f(\vec{r}) = \left[H(\vec{r}) + \gamma w(\vec{r})\right] O_\alpha f(\vec{r}) = 0. \tag{4.48}$$

From Eq. (4.48) it is again seen that if $f(\vec{r})$ is an eigenfunction of the operator of the wave equation, then so is $O_\alpha f(\vec{r})$. Consequently, in either of the systems in Eqs. (4.47) and (4.48), the eigenfunctions can be chosen to be eigenfunctions of both the wave equation and the set of point group symmetries. As shall be shown later, this is a great aid in classifying the solutions of the wave equation eigenvalue problems.

4.3   SYMMETRY GROUPS

One distinction between the translational symmetries of an infinite lattice and its point group symmetries is that while the translational symmetries of the lattice have an infinite number of different operations, the point groups involve a finite set of operations [1, 2, 4, 32, 33]. Both the translational symmetries and the point group symmetries, however, form a type of mathematical structure known as a group.

A group is a collection of operators which are linear transformations mapping space into itself. For example—translations, rotations, reflections, and inversions all map a space back into itself. Under the application of any of these operations each point in space is mapped to another point so that every space point is mapped by the operation and every space point is mapped onto by the operation. Furthermore, the set of symmetries forming the group includes the smallest collection of possible symmetries, and the successive application of any two symmetries of the set produces another symmetry operation of the set [1, 32, 33].

From Eq. (4.48) this means that

$$O_\alpha O_\beta \left[H(\vec{r}) + \gamma w(\vec{r})\right] f(\vec{r}) = \left[H(\vec{r}) + \gamma w(\vec{r})\right] O_\alpha O_\beta f(\vec{r}) = 0, \tag{4.49}$$

where [33]

$$O_\alpha O_\beta = O_\delta \tag{4.50}$$

such that O_α, O_β, O_δ are all members of the point group symmetries, and Eq. (4.49) is an equivalent statement to

$$O_\delta \left[H(\vec{r}) + \gamma w(\vec{r})\right] f(\vec{r}) = \left[H(\vec{r}) + \gamma w(\vec{r})\right] O_\delta f(\vec{r}) = 0. \tag{4.51}$$

In addition, it follows from the compositional properties of Eq. (4.49) that [1, 32, 33]

$$O_\alpha \left(O_\beta O_\gamma\right) = \left(O_\alpha O_\beta\right) O_\gamma, \tag{4.52}$$

that is, the order of the application of the operators is not important as long as their sequence in the product is maintained. This property is termed an associative property and is found in, e.g., the multiplications of reals and of integers.

Note that since the linear mappings forming the group [32, 33] are one-to-one mappings of space onto itself, an inverse to each operation of the group exists. In this regard, for each O_α of the set of group operations there is a single unique O_α^{-1} in the set such that

$$O_\alpha O_\alpha^{-1} = E, \tag{4.53}$$

where E is the identity operator leaving $f(\vec{r})$ unchanged under its application (i.e., it does nothing so that $Ef(\vec{r}) = f(\vec{r})$). The operations in Eqs. (4.49)–(4.53) from a complete definition of the group and its compositional operations.

It is important to note that unlike the arithmetic operation of the multiplication of real numbers, it was never indicated that the composition of two operators $O_\alpha O_\beta$ was equivalent to $O_\beta O_\alpha$. In general, this is not the case in symmetry group mappings [32, 33] and, consequently, it is often found that

$$O_\alpha O_\beta \neq O_\beta O_\alpha. \tag{4.54}$$

As a result, the composition of symmetry operations in Eq. (4.54) is known as group multiplication, and, unlike reals and integers, it is not communitive [32, 33].

Abelian and Non-Abelian Groups

In the case in which the group multiplication is communitive, the group is known as an Abelian group [32, 33]. It will be seen later that the statement that its symmetry group is Abelian has important consequences for the dispersion relations of the modes of a physical system. In particular, it leads to nondegenerate eigenvalues in the excitation spectra of the medium being studied.

An example of an Abelian group has already been encountered. The translational symmetry group of the crystal lattice is an Abelian group. As seen in Eqs. (4.8), (4.9), and (4.31) the various translations in the lattice can be composed from factors involving the smallest lattice translations, and the product of two lattice translations does not depend on the order in which the translations in the product are applied to a periodic function, $f(\vec{r})$.

An example of a non-Abelian group is provided by the rotational symmetries of a sphere. Here, in general, a rotation about the x-axis followed by a rotation about the y-axis is not equivalent to the same rotation about the y-axis followed by the same rotation about the x-axis. The final configurations of the system generated under these operations are found to be different for the two different procedures.

Both the rotation group (i.e., for isotropic space) and the translation group are infinite groups as their set of operations consists of an infinite number of distinct operations. The lattice point groups, however, are finite groups composed of sets containing a finite number of distinct operations. As an important point in this regard, it shall be shown later that the finite size of the point group of a physical system provides a fundamental limitation on the spectra of its modes of excitation.

Example of a Point Group

As an example of a point group, consider the set of symmetry operations of a square. This is a finite group, denoted as the C_{4v} point group [33, 34], containing eight symmetry operations mapping the square onto itself. The operations in C_{4v} represent the maximum set of point group symmetries found in a two-dimensional square lattice [33, 34] and will be important in later discussions of the modes of two-dimensional photonic and phononic crystals.

The eight symmetries of a square [32–34] contained in C_{4v} consist of: (1) the identity operation which leaves the square unchanged. This is denoted by E; (2) the anticlockwise rotations by $90°$, $180°$, $270°$ about a z-axis perpendicular to the plane of the square and passing through the center of the square. These are denoted as $R_4(i)$ for $i = 1, 2, 3$, respectively; and (3) the mirror reflections in the $x–z$ and $y–z$ planes where the x- and y-axes of the respective reflecting planes pass through the centers of opposite edges of the square. These operations are denoted as $R_2(1)$ and $R_2(2)$ for reflection through the vertical (i.e., y) and horizontal (i.e., x) axes, respectively. (4) The mirror reflection through the planes perpendicular to the plane of the square and respectively containing one of the two axes forming the diagonals of the square. These are denoted as $R'_2(1)$ and $R'_2(2)$ for the axis from the upper-right to the lower-left corners and the axis from the upper-left to the lower-right corners, respectively.

The multiplication table for these symmetry operations is written in Table 4.1. The table presents the results of the composition of two symmetry transformation as they act on a function $f(\vec{r})$. The first transformation is selected from the top row of the table and the second transformation is selected from the first column on the far left of the table.

As an example, choose $R'_2(2)$ from the top row of the table to form

$$R'_2(2) f(\vec{r}).$$ (4.55a)

This operation reflects the square about its diagonal running from its upper-left to its lower-right corners. Choosing $R_2(2)$ from the far-left column generates the transformation

$$R_2(2) R'_2(2) f(\vec{r})$$ (4.55b)

which reflects $R'_2(2) f(\vec{r})$ about the horizontal axis. From the table

$$R_2(2) R'_2(2) f(\vec{r}) = R_4(1) f(\vec{r})$$ (4.55c)

so that the product of the two rotations on the left is equivalent to a rotation by $90°$ about the axis perpendicular to the plane of the square and passing through its center.

From the table it is found that all product symmetries are equivalent to one of the single symmetries of the table. In addition, the rows and columns of the multiplication table contain each element of the group set once and only once. This is a characteristic of all groups, arising from the presence of an inverse element for every element of the group.

Table 4.1: The point symmetry group of a square denoted as C_{4v}

C_{4v}	E	$R_2(1)$	$R_2(2)$	$R'_2(1)$	$R'_2(2)$	$R_4(1)$	$R_4(2)$	$R_4(3)$
E	E	$R_2(1)$	$R_2(2)$	$R'_2(1)$	$R'_2(2)$	$R_4(1)$	$R_4(2)$	$R_4(3)$
$R_2(1)$	$R_2(1)$	E	$R_4(2)$	$R_4(1)$	$R_4(3)$	$R'_2(1)$	$R_2(2)$	$R'_2(2)$
$R_2(2)$	$R_2(2)$	$R_4(2)$	E	$R_4(3)$	$R_4(1)$	$R'_2(2)$	$R_2(1)$	$R'_2(1)$
$R'_2(1)$	$R'_2(1)$	$R_4(3)$	$R_4(1)$	E	$R_4(2)$	$R_2(2)$	$R'_2(2)$	$R_2(1)$
$R'_2(2)$	$R'_2(2)$	$R_4(1)$	$R_4(3)$	$R_4(2)$	E	$R_2(1)$	$R'_2(1)$	$R_2(2)$
$R_4(1)$	$R_4(1)$	$R'_2(2)$	$R'_2(1)$	$R_2(1)$	$R_2(2)$	$R_4(2)$	$R_4(3)$	E
$R_4(2)$	$R_4(2)$	$R_2(2)$	$R_2(1)$	$R'_2(2)$	$R'_2(1)$	$R_4(3)$	E	$R_4(1)$
$R_4(3)$	$R_4(3)$	$R'_2(1)$	$R'_2(2)$	$R_2(2)$	$R_2(1)$	E	$R_4(1)$	$R_4(2)$

4.3.1   SYMMETRIES AND EIGENVALUE DEGENERACIES

Earlier it was shown that translational symmetries played an important part in determining the general forms of the eigenvalue spectrum and of the eigenfunctions. In effect, the translational symmetry was found to be the origin of the stop and pass band structure of the dispersion relations of the modes of the system and of the formulation of the modal functions represented as planewaves multiplying spatial functions that are periodic in the crystal lattice. In addition, the regions in k−space containing a complete set of solutions for the modal dispersion relations and eigenfunctions was determined by the translational symmetry. Now it will be shown that the point group symmetries also have consequences for the eigenfunctions and eigenvalues of the system. For these symmetries, however, the effects are primarily found in the number of eigenfunctions which exist for a given eigenvalue of the system. These properties arising from the point group symmetries are closely related to the structure of the point group multiplication tables [33].

In the case of the wave equation eigenvalue problem, the eigenvalues can have one or more linearly independent eigenfunctions associated with them. The number of eigenfunctions corresponding to an eigenvalue has important consequences for the orthogonality properties of the system as well as on the effects of symmetry breaking perturbations on the system. When more than one linearly independent eigenfunctions exists for an eigenvalue, the eigenvalue is said to be degenerate, and this degeneracy has consequences for the orthogonality of the modes. Only modes of different eigenvalues are guaranteed to be orthogonal so that care must be taken with the linearly independent solutions arising from a degenerate eigenvalue [1, 32–34]. It is, however, a simple matter to create an orthogonal set of eigenfunctions from the linearly independent set of modes associated with a degenerate eigenvalue. This is usually done using Gram–Schmidt methods [32] and leads to a complete set of eigenfunctions in which to study the wave equation problem.

The degeneracy of an eigenvalue in general arises from the existence of point group symmetries in the system being studied and the structuring of these symmetries in the group multiplication table [1, 32–34]. In this regard, a particularly important feature needed to understand the properties of the modal dispersion relations is the classifications of the types of symmetry operations in the point group into sets of like and unlike symmetry operations. The nature of the eigenvalue degeneracies, in fact, is obtained directly from the classification of the group elements in group multiplication tables, such as Table 4.1, into sets composed of similar symmetry operations. This classification, involving the partitioning of the group elements into sets known as classes, is now discussed.

4.3.2   CLASSES OF THE GROUP

The elements of a group multiplication table can be partitioned in to sets, each set containing only elements of similar type of symmetry operations. By similar symmetry operations we mean operations which are of the same type of symmetry but which act about different points, planes, or axes of the system being studied. In this partition, the sets of the partition are referred to as classes and each element of the group occurs in one and only one of the partitioning sets or classes.

For example, consider the set of symmetry operations of a cube. One class of these symmetries is composed from the set of rotations by 180° about each of the three axes through the center of opposite faces of the cube. This forms a class of three distinct elements of the group. Another class contains the set of rotations by ±90° about the same axes. (Note: The distinction between these two classes is in the angle of rotation.) Here the rotational axes of the cube bear the same relationship to the cube geometry but are different axes in space. As another example from the symmetry of the cube, the set of rotations by ±120° about the three diagonal axes passing through the center of the cube are a distinct class. As a final point, it is interesting to note that in all groups, the identity, E, which leaves the object unchanged is in a class by itself.

In general, it can be shown in the partitioning of a group into classes that, if

$$A = C^{-1}BC \tag{4.56}$$

for the symmetry element A, B, C of a group, the symmetry elements A and B are in the same class [32, 33]. In this way, the similarity transform defined in Eq. (4.56) can be used to partition all the elements of a group uniquely into classes.

As an example which will be useful in the study of two-dimensional photonic and phononic crystals, consider the C_{4v} point symmetries of the square lattice given in Table 4.1 and their partitioning into classes. In the table the elements $\{R_2(1), R_2(2)\}$ is a class representing mirror reflections through planes containing the vertical and horizontal axes. They are seen to be the same types of operations, similarly performed on the geometry of the square. The set $\{R_2(1), R_2(2)\}$ is a class which we will denote as σ_v. Similarly, the elements $\{R'_2(1), R'_2(2)\}$ represent mirror reflections in planes containing the diagonals of the square.

The set $\{R'_2(1), R'_2(2)\}$ is a class which we will denote as σ_d. The identity element E is unique as it is the only element of the group which does nothing to the square. The identity is the sole element of a class which will be denoted as $\{E\}$. Similarly, the element $\{R_4(2)\}$, representing a rotation by 180° about the axis perpendicular to the plane of the square, is a class composed of a unique type of symmetry in the group. The class $\{R_4(2)\}$ is denoted C_2. The remaining elements $\{R_4(1), R_4(3)\}$, representing a rotation by ±90° about the axis perpendicular to the plane of the square, are rotations about the same axis but in different direction. The class $\{R_4(1), R_4(3)\}$ is denoted by C_4. The complete partitioning of the group into classes is given by: $\{E\}$, $\{R_4(2)\}$, $\{R_4(1), R_4(3)\}$, $\{R_2(1), R_2(2)\}$, $\{R'_2(1), R'_2(2)\}$.

4.3.3   IMPORTANCE OF THE CLASSES

To see how the classes are important [1, 32–34] in understanding the nature of the eigenvalue degeneracies consider the eigenvalue problem in Eq. (4.23). Let $\left\{\varphi_i^{(n,m)}(\vec{r})\right\}$ be the normalized eigenfunctions of Eq. (4.23) corresponding to the set of eigenvalues $\left\{\gamma_i^{(n)}\right\}$. With this notation the eigenvalue problem reads

$$\left[H(\vec{r}) + \gamma_i^{(n)} w(\vec{r})\right] \varphi_i^{(n,m)}(\vec{r}) = 0 \quad \text{for} \quad m = 1, 2, 3, \ldots, n. \tag{4.57a}$$

Here i labels the ith eigenvalue of the system and its eigenfunctions, n indicates the number of degenerate eigenfunctions corresponding to $\gamma_i^{(n)}$, $\varphi_i^{(n,m)}(\vec{r})$ is any one of the degenerate eigenfunctions of $\gamma_i^{(n)}$, and $m = 1, 2, 3, \ldots, n$ labels the n different (linearly independent and orthonormalized) eigenfunctions corresponding to $\gamma_i^{(n)}$. In this notation, $\gamma_i^{(n)}$ is a degenerate eigenvalue associated with n degenerate eigenfunctions, and each of the n different eigenfunctions of $\gamma_i^{(n)}$ is labeled by $m = 1, 2, 3, \ldots, n$.

For the case in which $H(\vec{r})$ is a Hermitian operator, the eigenvalues in Eq. (4.57a) are real, and the eigenfunctions corresponding to different eigenvalues are orthogonal. Specifically, for the normalized eigenfunctions $\varphi_i^{(n,m)}(\vec{r})$ and $\varphi_{i'}^{(n',m')}(\vec{r})$ corresponding to eigenvalues $\gamma_i^{(n)}$ and $\gamma_{i'}^{(n')}$, respectively, the orthogonality relations between the two functions are

$$\int d\vec{r}\, \varphi_{i'}^{*(n',m')}(\vec{r})\, \varphi_i^{(n,m)}(\vec{r})\, w(\vec{r}) = \delta_{i,i'}\delta_{n,n'}\delta_{m,m'}. \tag{4.57b}$$

Here it is assumed that the set of degenerate eigenfunctions have also been orthonormalized using a Gram–Schmidt procedure.

From Eq. (4.48) it is seen that if $\varphi_i^{(n,m)}(\vec{r})$ is an eigenfunction of the problem with eigenvalue $\gamma_i^{(n)}$, then so is $O_\alpha \varphi_i^{(n,m)}(\vec{r})$ where O_α is one of the elements in the symmetry group $\{O_\alpha\}$ of the wave equation. Consequently, the entire set of functions $\left\{O_\alpha \varphi_i^{(n,m)}(\vec{r})\right\}$ are eigenfunctions of the eigenvalue $\gamma_i^{(n)}$. It then follows that O_α acting in the set of $\varphi_i^{(n,m)}(\vec{r})$ can be represented by the linear transformation [33]

$$O_\alpha \varphi_i^{(n,m)} (\vec{r}) = \sum_{p=1}^{n} \varphi_i^{(n,p)} (\vec{r}) \, U_{p,m}^{(n,i)} (\alpha) \tag{4.58a}$$

where

$$U_{p,m}^{(n,i)} (\alpha) \quad \text{for} \quad m, p = 1, 2, 3, \dots, n \tag{4.58b}$$

is an $n \times n$ unitary matrix characterizing the application of O_α on $\varphi_i^{(n,m)} (\vec{r})$. The results of the actions of the symmetry operator are, consequently, represented in terms of the $\left\{ \varphi_i^{(n,m)} (\vec{r}) \right\}$ forming a complete set of states in the space of the degenerate eigenfunctions of $\gamma_i^{(n)}$. In this regard, from Eq. (4.58) it is seen that the transformation of $\varphi_i^{(n,m)}$ under the actions of the group of symmetry transformations maps the space $\left\{ \varphi_i^{(n,m)} (\vec{r}) \right\}$ associated with the $\gamma_i^{(n)}$ back onto itself.

The eigenvalue problem in Eq. (4.57) and the set of symmetries $\{O_\alpha\}$ can be expressed as matrices formulated in the basis of the eigenfunctions from Eq. (4.57). In this basis the problem in Eq. (4.57) takes the simple form of a diagonal matrix while the matrices of $\{O_\alpha\}$ become block diagonal matrices. To see this, multiply the eigenvalue problem in Eq. (4.57a) by $\varphi_j^{*(l,p)} (\vec{r})$ and integrate. The integration gives the set of relationships

$$\int d\vec{r} \varphi_{i'}^{*(n',m')} (\vec{r}) \, H (\vec{r}) \, \varphi_i^{(n,m)} (\vec{r}) = H_{i',i}^{n',n;m',m}$$
$$= -\gamma_i^{(n)} \int d\vec{r} \varphi_{i'}^{*(n',m')} (\vec{r}) \, \varphi_i^{(n,m)} (\vec{r}) \, w (\vec{r}) = -\gamma_i^{(n)} \delta_{i,i'} \delta_{n,n'} \delta_{m,m'}, \tag{4.59a}$$

yielding the matrix equation

$$H_{i',i}^{n',n;m',m} = -\gamma_i^{(n)} \delta_{i,i'} \delta_{n,n'} \delta_{m,m'}. \tag{4.59b}$$

In the representation in Eq. (4.59b) the eigenvalue matrix is a diagonal matrix with the eigenvalue $\gamma_i^{(n)}$ occurring n times on the diagonal for each of the n degenerate eigenfunctions, $\left\{ \varphi_i^{(n,m)} (\vec{r}) \right\}$, corresponding to $\gamma_i^{(n)}$. Figure 4.4 shows a representation of the diagonal matrix of eigenvalues written in this representation. In this case the A_i would be diagonal matrices containing the same eigenvalues on their diagonal.

The representation of the set of symmetries $\{O_\alpha\}$ as matrices in the basis of the eigenfunctions of Eq. (4.57) is a little more complicated. From Eq. (4.58) it follows that under the action of any of the O_α in the group, a degenerate eigenfunction corresponding to a given $\gamma_i^{(n)}$ maps into a linear combination of the entire set of degenerated eigenfunctions of $\gamma_i^{(n)}$. Consequently, if the basis eigenfunctions are organized into groupings of mutually degenerate eigenfunctions,

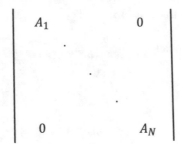

Figure 4.4: Block diagonal form. The square matrices $A_1, \ldots, A_n$ occur along the diagonal of the large (block diagonal) matrix and are composed of elements with arbitrary numerical values. The other elements of the large matrix (i.e., outside the matrices $A_1, \ldots, A_n$) are all zero. The $n \times n$ block A_i corresponds the degenerate eigenvalue $\gamma_i^{(n)}$.

the matrices representing $\{O_\alpha\}$ in the ordered basis form block diagonal matrices. The blocks of each matrix represent the transformation of the degenerate eigenfunctions of that block among themselves. This is illustrated in Fig. 4.4, showing a representation of a block diagonal matrix in which each of the different blocks correspond to a different $\gamma_i^{(n)}$. That is, now A_i in Fig. 4.4 is a block matrix corresponding to the eigenvalue $\gamma_i^{(n)}$.

The block diagonal matrices representing the $\{O_\alpha\}$ symmetries all have the same block diagonal form and reproduce the group multiplication tables when multiplied together under normal matrix multiplications. This follows because the matrices represent linear transformations and, as noted earlier, the group multiplication table represents the actions of symmetry operations mapping a space onto itself. Furthermore, any set of blocks corresponding to the same degenerate eigenfunctions $\left\{\varphi_i^{(n,m)}(\vec{r})\right\}$ of $\gamma_i^{(n)}$ also, under matrix multiplication, represent the $\{O_\alpha\}$ symmetries. The representation of the group by the set of blocks in Eq. (4.58) formed from the degenerate eigenfunctions of $\gamma_i^{(n)}$ is known as an irreducible representation, and the representation formed from all of the eigenfunctions of the problem is known as a reducible representation. In this way the reducible representation is composed of a block diagonal arrangement of the irreducible representations.

The importance of the distinction between the reducible and the irreducible representations is that the irreducible representations are related to the degeneracies of the eigenvalues of the system while a reducible representation is related to the number of occurrences of these degeneracies in the system. In this regard, from Eq. (4.58), it is seen that the dimension of the matrices of the irreducible representation of the group for the degenerate eigenvalue $\gamma_i^{(n)}$ equals the number of orthonormal eigenfunctions of $\gamma_i^{(n)}$.

A reducible representation then is composed of the sets of irreducible representations that are possible in the group. It will now be seen that the dimensions of the irreducible representations possible in the group can be determined from the group multiplication table. This is a great simplification in the determination of the degeneracies possible in a physical system.

The object in the following discussions is to understand the degeneracies of the eigenvalues using the group multiplication table. As the detailed mathematics needed to develop the theory of this topic is complex and unnecessary to a knowledge of the application of the theory, it is only outlined here. For a more detailed treatment of the development of the theory the reader is referred to the literature [1, 32, 33].

4.3.4 IRREDUCIBLE REPRESENTATIONS AND CLASS ORTHOGONALITY RELATIONSHIPS

An irreducible representation of the symmetry group of the eigenvalue problem represents a transformation of the eigenfunctions of the degenerate eigenvalue, $\gamma_i^{(n)}$, into themselves under the symmetry group of the problem [1, 32–34]. Since the irreducible representation is a set of finite $n \times n$ matrices which reproduce the point group multiplication table under matrix multiplication, the matrices of the representation are highly restricted in their form.

For example, let $\{M_\alpha\}$ be a set of unitary matrices of an irreducible representation of the group of symmetry operations $\{O_\alpha\}$, then the composition of the symmetry operations $O_\alpha O_\beta = O_\delta$ is represented by the matrix multiplication $M_\alpha M_\beta = M_\delta$. For this representation to hold for all of the symmetry compositions required in the group multiplication table, the matrices must be of highly specific forms. Fortunately, it can be shown that a knowledge of the detailed structure of the matrices in the irreducible representations is not needed, and only a knowledge of the traces of the matrices is needed to determine the nature of the eigenvalue spectrum.

The traces of the matrices $\{M_\alpha\}$ are an invariant property of each matrix, being independent of the basis set used to compose the matrix. This follows due to the invariance of the trace under a similarity transformation, i.e., for nonsingular matrices A and C it follows that

$$Tr\ A = Tr\ C^{-1}AC. \tag{4.60a}$$

In addition, from this invariance it is readily seen that if three nonsingular matrices are related by

$$A = C^{-1}BC, \tag{4.60b}$$

then $Tr\ A = Tr\ B$. Consequently, since matrices of $\{M_\alpha\}$ representing symmetries in the same class are related to one another through similarity transformations they must all have the same trace [1, 32–34].

The traces of the irreducible representations play an important part in the following discussions, and because of this they have been specifically termed as "characters" of the matrices and of the classes the matrices compose. In this regard, for the group of matrices $\{M_\alpha\}$ formed under matrix multiplication the character of the matrix M_β in this group is denoted by

$$\chi\left(M_\beta\right) = Tr\left(M_\beta\right) \tag{4.60c}$$

Table 4.2: The character table for C_{4v}

C_{4v}	E	C_2	$2C_4$	$2\sigma_v$	$2\sigma_d$
A_1	1	1	1	1	1
A_2	1	1	1	-1	-1
B_1	1	1	-1	1	-1
B_2	1	1	-1	-1	1
E	2	-2	0	0	0

so that the set of characters $\{\chi(M_\alpha)\}$ provide a characterization of the matrices in the group $\{M_\alpha\}$.

When the group is partitioned into classes, a class C_α is composed as the set of irreducible matrices $C_\alpha = \{M_1, M_2, \ldots, M_j\}$ belonging as a subset to the group $\{M_\alpha\}$. In this subset all of the matrices are related to one another by a similarity transformation as in Eq. (4.60b) and share a class character

$$\chi(C_\alpha) = Tr(M_1) = Tr(M_2) = \ldots = Tr(M_j) \tag{4.60d}$$

which is common to all of the matrices in the class. In these representations the properties of the matrices and classes are summarized by a single number [33].

In the following, we shall see that a knowledge of the classes and the traces of the irreducible representations in the various classes is very important to a determination of the eigenvalue degeneracies present in a system [1, 32–34]. Due to this importance, the traces of the matrices in the irreducible representations of the various crystallographic point group symmetries have been worked out and are commonly presented in published tabular form [33]. The tabulation generally lists the name of the crystallographic point symmetry group, the names of the irreducible representations of the symmetry group, and the values of the characters listed for each class in the group.

An Example

An example of such a tabulation for the point group symmetries of a square is listed in Table 4.2. The group has been termed by crystallographers as C_{4v} and was introduced earlier in Table 4.1. Some of the standard notation used in the tabulation of Table 4.2 is now discussed.

The top row of the table lists the classes: E, C_2, C_4, σ_v, σ_d where $E = \{E\}$ $C_2 = \{R_4(2)\}$, $C_4 = \{R_4(1), R_4(3)\}$, $\sigma_v = \{R_2(1), R_2(2)\}, \sigma_d = \{R'_2(1), R'_2(2)\}$. The 2 prefixed on C_4, σ_v, σ_d in the top row indicates that there are two group elements in these classes, while the classes E, C_2 (prefixed 1) have only a single element. The left-most column contains the names of the five different irreducible representations in a crystallographic notation. The rest of the table contains the numerical values of the characters $\chi(E)$, $\chi(C_2)$, $\chi(C_4)$, $\chi(\sigma_v)$, $\chi(\sigma_d)$ listed for each of the irreducible representations A_1, A_2, B_1, B_2, E.

An interesting point of the classification of the group elements into classes is that the number of different classes is equal to the number of distinct irreducible representations of the group. This is true of irreducible matrix representations in general. In addition, the column labeled by the identity representation E contains only the identity matrix as an element. Consequently, the trace or character of the identity matrix is equal to its dimension and that of the other matrices in its particular irreducible representation.

In this regard, the representations A_1, A_2, B_1, B_2 are all 1×1 matrices, i.e., the matrices are just single numbers. An earlier example of a 1×1 matrix representation was encountered (i.e., see Eq. (4.9)–(4.11)) in the representations of the group of translational symmetries. There it was found that the group was represented by the phases $\sigma = e^{-ika}$ generated by the translation of the wavefunctions along a space lattice. The translation group was an Abelian group, and it can be shown that all Abelian groups (of which the translation group is a particular example) are represented by 1×1 matrices.

On the other hand, the irreducible representation labeled E is seen from Table 4.2 to be two-dimensional. This results from the non-commutative nature of the group C_{4v} as only Abelian groups solely have one-dimensional representations. In addition, to these basic properties of the representations of the general structure of the group and its partition into classes, there are a series of interesting orthogonality properties involving the rows and columns of the table of characters themselves. Later, these are shown to be of fundamental importance in determining the eigenvalue degeneracies of a physical system. The orthogonality properties are now discussed.

Orthogonality of the Characters

The table of characters in Table 4.2 illustrates some of the interesting orthogonality properties of the rows and columns of the characters. In general, it is found that for the rows of the character table

$$\sum_k \chi^{(i)}(k)^* \chi^{(j)}(k) N_k = h \delta_{i,j}, \tag{4.61}$$

where $\chi^{(i)}(k)$ is the character of the ith irreducible representations (e.g., $i = A_1$, A_2, B_1, B_2, E for the square in Table 4.2), k is one of the classes (e.g., E, C_2, C_4, σ_v, σ_d for the square in Table 4.2), N_k is the prefix integers in the listing of classes in the top row of the table, and h is the number of elements in the group. It is also found that for the columns of the character table

$$\sum_i \chi^{(i)}(k)^* \chi^{(i)}(l) = \frac{h}{N_k} \delta_{k,l}. \tag{4.62}$$

These two orthogonality relations are readily seen to be obeyed by the characters in Table 4.2.

4.4 EXAMPLE OF THE CHARACTER ORTHOGONALITY RELATIONSHIPS

The orthogonally relations in Eqs. (4.61) and (4.62) are very important in that they can be used to determine the number of irreducible representations that are contained within a general reducible representation of the symmetry group. This follows as ultimately every matrix representation of a symmetry group can be placed into the form of a set of block diagonal matrices. In this block diagonal form of the matrix representation the blocks are formed from the irreducible representations of the group [1, 32–34].

To see an example of this, consider a two-dimensional reducible representation of the group C_{4v} which is composed from the A_1 and A_2 one-dimensional matrix representations. A particular set of two by two block diagonal matrices which provides such a representation of C_{4v} under matrix multiplication is

$$\left\{ E = \begin{vmatrix} 1 & 0 \\ 0 & 1 \end{vmatrix}, \quad R_4(2) = \begin{vmatrix} 1 & 0 \\ 0 & 1 \end{vmatrix}, \quad R_4(1) = R_4(3) = \begin{vmatrix} 1 & 0 \\ 0 & 1 \end{vmatrix}, \right.$$
$$\left. R_2(1) = R_2(2) = \begin{vmatrix} 1 & 0 \\ 0 & -1 \end{vmatrix}, \quad R'_2(1) = R'_2(2) = \begin{vmatrix} 1 & 0 \\ 0 & -1 \end{vmatrix} \right\}. \tag{4.63}$$

The set of matrices in Eq. (4.63) have the A_1 representation in the upper-left block and the A_2 representation in its lower-right block. These blocks separately and independently represent the group multiplication table as do the two by two matrices they compose in Eq. (4.63). (Notice, however, that though the matrices in Eq. (4.63) are block diagonal in A_1 and A_2, a similarity transformation applied to all of the matrices in Eq. (4.63) can take them out of diagonal form. This transformation represents a change of the basis in which the matrices are expressed but does not change the matrix characters.)

For the matrices of the reducible representation in Eq. (4.63) the basis-independent characters are

$$\{\chi(E) = 2, \quad \chi(C_2) = 2, \quad \chi(C_4) = 2, \quad \chi(\sigma_v) = 0, \quad \chi(\sigma_d) = 0\}. \tag{4.64}$$

From these characters of the reducible representation it is a simple matter to determine what are the irreducible representations which compose the reducible representation and how many times each irreducible representation enters into the reducible representation. This determination follows from an application of the orthogonality relation in Eq. (4.61) to project out from the characters in Eq. (4.64) each of the irreducible representations composing the reducible representation in Eq. (4.63).

To see this, consider the characters in Eq. (4.64) and use the relationship in Eq. (4.61) to project these onto the characters of the irreducible representations in the rows of Table 4.2.

Applying the orthogonality relationship in Eqs. (4.61)–(4.64) it follows that

$$\sum_k \chi^{(A_1)}(k)^* \chi(k) N_k = 8 \tag{4.65a}$$

projects out the irreducible representation of A_1. The 8 on the right-hand side indicates that A_1 occurs in the reducible representation once. If the right-hand side were $8n$ then A_1 would occur n times in the reducible representation.

Similarly,

$$\sum_k \chi^{(A_2)}(k)^* \chi(k) N_k = 8 \tag{4.65b}$$

projects out the irreducible representation of A_2 once. For the other representations in Table 4.2

$$\sum_k \chi^{(i)}(k)^* \chi(k) N_k = 0 \quad \text{for} \quad i = B_1, \, B_2, \, E, \tag{4.65c}$$

which indicates an absence of these representations in the representation in Eqs. (4.63) and (4.64).

Consequently, the row orthogonality relationships indicate that there is one A_1 irreducible representation and one A_2 irreducible representation in the reducible representation of Eq. (4.63). None of the other irreducible representations is present in Eq. (4.63).

The rows of characters in Table 4.2 form a complete orthogonal set in the five-dimensional character space. All of the reducible representations are decomposed in this space into multiple sets of irreducible representations in this space.

While the orthogonality condition in Eq. (4.61) is useful in projecting out the various irreducible representations contained in a reducible representation, some important information about the degeneracies of an eigenvalue spectrum can be obtained by just looking at the character table for the symmetry group of a physical system. The following discussion will focus on these aspects of the group theory.

4.5   AN APPLICATION OF THE CHARACTER TABLE IN BAND STRUCTURE CALCULATIONS

One of the applications of group representations and character tables is in the determination of the degeneracies of the modal eigenvalues in systems exhibiting point group symmetry [33, 34]. In crystal structures this involves studying the possible degeneracies of the eigenvalues and how these degeneracies change as the wavevector of the modal eigenfunctions change in wavevector space. In the following an example of such an application is given for the study of the band structure of a crystalline system represented by a Helmholtz equation defined on a square lattice. The results of this study are important in understanding the properties of many types of two-dimensional photonic and phononic crystals.

In particular, consider the two-dimensional Helmholtz equation of the form

$$\nabla^2 F\left(\vec{r}\right) + \frac{\omega^2}{c^2}\varepsilon\left(\vec{r}\right) F\left(\vec{r}\right) = 0, \tag{4.66}$$

where $\vec{r}$ is a position vector in the x_1–x_2 plane and $\varepsilon\left(\vec{r}\right)$ is a periodic function defined over a square lattice with lattice constant a. This provides a model of a photonic crystal, but by re-defining the periodic term $\frac{\omega^2}{c^2}\varepsilon\left(\vec{r}\right)$, can also be made to represent a phononic crystal. For this application and the symmetry discussions which follow, the function $F\left(\vec{r}\right)$ is treated as a scalar field, e.g., offering considerations of systems with electric or magnetic fields polarized parallel to an array of dielectric cylinders.

From the earlier discussions on the eigenfunctions of periodic systems, the form of $F\left(\vec{r}\right)$ may be represented by

$$F\left(\vec{r}\right) = e^{i\vec{k}\cdot\vec{r}}u_{\vec{k}}\left(\vec{r}\right), \tag{4.67}$$

where $u_{\vec{k}}\left(\vec{r}\right)$ is periodic over the square lattice in the x_1–x_2 plane and

$$\vec{k} = \frac{\pi}{Na}\left(n\hat{x}_1 + m\hat{x}_2\right) \quad \text{with} \quad m, n = 0, \pm1, \pm2, \ldots., \pm N \tag{4.68}$$

for a $2N \times 2N$ lattice as $N \to \infty$. The eigenfunctions of Eq. (4.66) are consequently character-ized as modes of the wavevector $\vec{k}$.

In Fig. 4.5, the square zone in $k-$ or wavevector space centered about the origin and containing all of the unique solutions of the square lattice Helmholtz equation is represented. The outlined square encloses the wave vectors $-\frac{\pi}{a} < k_1, k_2 < \frac{\pi}{a}$, and corresponding to each of these wavevectors are a number of different frequency solutions associated with the different bands of the modal dispersion relation. In addition, a number of symmetry lines and points in this region of $k-$ space have been indicated and label Γ, Δ, X, Z, M, Σ. These features are provided in order to facilitate the symmetry discussions which follow.

In the following, a group theory treatment regarding the degeneracies of the eigenvalues of the Helmholtz modal solutions is given for the modes which exist along the line $\overline{\Gamma\Delta X}$. As the wavevector of the solutions changes along the line the nature of the degeneracy of the solutions is found to change. This change can be understood in terms of the changes of the character tables of the symmetry groups describing the symmetry of the system along the line. The change in symmetry of the solutions is found to be directly related to the wavevectors of the solutions along this line. Similar arguments to those presented below can be made for other lines of symmetry of the system studied, and it is left to the reader to continue the present discussions to these other lines of symmetry in the wavevector space of the square lattice.

Upon substituting Eq. (4.67) into Eq. (4.66), it follows that at a general wavevector within the square in Fig. 4.5 the equation [1, 2, 32–34]

$$\left[\nabla^2 + i\vec{k}\cdot\nabla - k^2 + \frac{\omega^2}{c^2}\varepsilon\left(\vec{r}\right)\right]u_{\vec{k}}\left(\vec{r}\right) = 0 \tag{4.69}$$

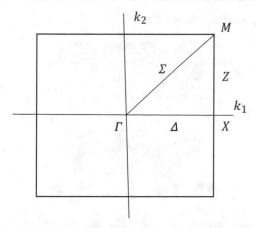

Figure 4.5: The square zone of unique solutions in wavevector space for the Helmholtz equations on the square lattice. The origin of wavevector space is located in the center of the square at the point Γ.

determines the periodic part of $F(\vec{r})$. It is readily seen that the equation is highly dependent on $\vec{k}$ and that the direction of $\vec{k}$ is special in space. From the form of Eq. (4.69) it follows that as the vector $\vec{k}$ is changed from zero and passes through various nonzero values, the symmetry properties of Eq. (4.69) are changed. This dependence, in fact, represents the interaction of the translational symmetry group with the point group symmetries of the model.

As discussed in Eq. (4.60), the degeneracies of the eigenvalues $\frac{\omega^2}{c^2}$ at a given $\vec{k}$ are now shown to be closely related to the properties of the characters of the point group symmetries at $\vec{k}$. These point group symmetries are influenced to a great extent by the presence of the wavevectors $\vec{k}$ as a fixed feature in Eq. (4.69), and significant changes are found in the eigenvalue spectrum as the characters of the system change from point to point throughout wavevector space.

Regarding Eq. (4.69), first consider the case of the solution at the point Γ in Fig. 4.5. At this point $\vec{k} = 0$, and the equation for $u_{\vec{k}=0}(\vec{r})$, consequently, takes the form

$$\left[\nabla^2 + \frac{\omega^2}{c^2} \varepsilon(\vec{r}) \right] u_{\vec{k}}(\vec{r}) = 0. \tag{4.70}$$

In the absence of the symmetry breaking $\vec{k} \neq 0$ terms in Eq. (4.69), the symmetry group of Eq. (4.70) is composed of the translational symmetries and the square point group symmetries of Table 4.1. From the characters of the square point group in Table 4.2, it is evident that there are only one-dimensional and two-dimensional irreducible representations in the point group. Consequently, the eigenvalue degeneracies of Eq. (4.70) are either non-degenerate eigenvalues or doubly degenerate eigenvalues [32–34]. Higher degeneracies in the eigenvalue spectrum are not possible with the square lattice point group. For the case considered here, Eq. (4.70)

Table 4.3: C_{1h} point group symmetries of Eq. (4.71) and the C_{1h} character table

C_{1h}	E	$R_2(2)$
E	E	$R_2(2)$
$R_2(2)$	$R_2(2)$	E

C_{1h}	E	σ
A	1	1
B	1	-1

represents the highest degree of point group symmetry possible in the system represented by Eq. (4.69).

Next consider the case of points along the line labeled by Δ in Fig. 4.5. Along this line the points are of the general form $k_1 \neq 0$ and $k_2 = 0$, and Eq. (4.69) now becomes

$$\left[\nabla^2 + i k_1 \frac{\partial}{\partial x_1} - k_1^2 + \frac{\omega^2}{c^2} \varepsilon (\vec{r}) \right] u_{\vec{k}} (\vec{r}) = 0. \tag{4.71}$$

This represents a case of Eq. (4.69) with much less symmetry than that considered in Eq. (4.70). Due to the x_1 dependence in Eq. (4.71) the equation no longer exhibits the C_{4v} square lattice point group symmetries. The new point group symmetries (denoted as the C_{1h} point group) of Eq. (4.71) are listed in Table 4.3 and consist of the identity operation and a mirror reflection symmetry about the horizontal axis which is given by the operation $R_2 (2)$. Here the notation used in Tables 4.1 and 4.2 are carried over in both the group and character tables.

It is found that the system described by Eq. (4.71) only has two irreducible representations named A and B. Both of these are one-dimensional irreducible representations so that the eigenvalues of Eq. (4.71) are only non-degenerate. In the following, it shall be examined how eigenvalues of Eq. (4.70) for the square lattice symmetry transition to the eigenvalues of the system described by Eq. (4.71) with the change in $\vec{k}$. This involves a comparison of the character tables of the two systems in Eqs. (4.70) and (4.71).

In the comparison of Eqs. (4.70) and (4.71) the symmetry operations common to the two groups are $\{E, \ R_2 (2)\}$. For both groups these elements are represented by the character E and a single character of the class σ_v in C_{4v} corresponding to the character σ in C_{1h}, respectively. In this regard, consider the characters of the classes E and $\sigma_v = \sigma$ in the two different group character tables. Results for these characters are taken from Table 4.2 for the square point group of Eq. (4.70) and from Table 4.3 for the reduced symmetry in Eq. (4.71) and are listed in Table 4.4 labeled by the representations $E, \ A, \ B$ from which they come. Here E is from the two-dimensional representation in Table 4.2, and A and B are from the character table in Table 4.3.

Table 4.4: Table of the E and $\sigma_v = \sigma$ characters of the group elements E and R_2 (2)

	E	$\sigma_v = \sigma$
A	1	1
B	1	-1
E	2	0

From a comparison, it is readily seen in Table 4.4 that summing the columns of A and B gives the characters of the two-dimensional representation E. (More precisely, the row orthogonality relations of the character table in Table 4.3 has been used to project these irreducible representations from the E representation in Table 4.2.) This indicates that upon transitioning from Eqs. (4.70) to (4.71) the irreducible representations A and B are both contained once within the now reducible representation E. Consequently, leaving $\vec{k} = 0$ and traveling along $k_1 \neq 0$ the degeneracy of the eigenvalues associated with E are split and become non-degenerated.

Finally, consider the special point at X in Fig. 4.5. This point is at the right-hand edge of the set of unique solutions at $k_1 = \frac{\pi}{a}$ and $k_2 = 0$ and is an end point along the line $k_1 \neq 0$ and $k_2 = 0$ of solutions. (Remember, here a is the nearest neighbor separation on the square lattice.) The form of Eq. (4.69) now becomes

$$\left[\nabla^2 + i \frac{\pi}{a} \frac{\partial}{\partial x_1} - \frac{\pi^2}{a^2} + \frac{\omega^2}{c^2} \varepsilon \left(\vec{r} \right) \right] u_{\vec{k} = \frac{\pi}{a} \hat{x}_1} \left(\vec{r} \right) = 0, \qquad (4.72a)$$

which, due to the $\frac{2\pi}{a}$ periodicity of the solutions in wavevector space, has the same solution set as

$$\left[\nabla^2 - i \frac{\pi}{a} \frac{\partial}{\partial x_1} - \frac{\pi^2}{a^2} + \frac{\omega^2}{c^2} \varepsilon \left(\vec{r} \right) \right] u_{\vec{k} = -\frac{\pi}{a} \hat{x}_1} \left(\vec{r} \right) = 0. \qquad (4.72b)$$

The wave function solutions of Eq. (4.72) differ from those of Eqs. (4.70) and (4.71) in that they are standing waves (not planewaves) formed from an equal mixture of left and right propagating waves along the x_1 direction. In this regard, the two points on the left and right edges of the square located on the k_1-axis are actually the same point [32–34].

Again, due to the x_1 dependence in Eq. (4.72) the equation no longer exhibits the square lattice point group symmetries, but now the points $\vec{k} = \frac{\pi}{a} \hat{x}_1$ and $\vec{k} = -\frac{\pi}{a} \hat{x}_1$ represent the same solutions. This identification opens up additional symmetries in the system from those listed in Table 4.3 for the general points along the line $k_1 \neq 0$ and $k_2 = 0$. The new symmetry group of the system, however, is Abelian so that the irreducible representations of the new symmetry group are one-dimensional.

Consequently, the new point group symmetries (denoted as the symmetries of the C_{2v} point group) of Eq. (4.72) are listed in Table 4.5. Here the notation used in Tables 4.1 and

Table 4.5: C_{2v} point group symmetries of Eq. (4.72) and the C_{2v} character table

C_{2v}	E	$R_2(2)$	$R_4(2)$	$R_2(1)$
E	E	$R_2(2)$	$R_4(2)$	$R_2(1)$
$R_2(2)$	$R_2(2)$	E	$R_2(1)$	$R_4(2)$
$R_4(2)$	$R_4(2)$	$R_2(1)$	E	$R_2(2)$
$R_2(1)$	$R_2(1)$	$R_4(2)$	$R_2(2)$	E

C_{2v}	E	C_2	$\sigma(xz)$	$\sigma(yz)$
A_1	1	1	1	1
A_2	1	1	-1	-1
B_1	1	-1	1	-1
B_2	1	-1	-1	1

4.2 are carried over in both the group and character tables, and the group multiplication table has more symmetry elements than that in Table 4.3 but less than that in Tables 4.1 and 4.2. In addition, the character table for the group multiplication table in Table 4.5 has more classes than that in Table 4.3.

The point symmetry group for Eq. (4.72) is now found to have four irreducible representations named A_1, A_2, B_1, and B_2. All of the four irreducible representations are one-dimensional so that the eigenvalues of Eq. (4.72) are all non-degenerate. Each of the classes contain only one group element such that the class E contains the identity; the class $\sigma(xz)$ contains the mirror reflection about the horizontal, $R_2(2)$; the class C_2 contains the 180° rotation about the axis perpendicular to the page; and the class $\sigma(yz)$ contains the mirror reflection about the vertical axis of the square.

To see how the irreducible representations A_1, A_2, B_1, and B_2 are transformed upon leaving the point X at $k_1 = \frac{\pi}{a}$ and $k_2 = 0$ and moving along the line labeled Δ for which $k_1 \neq 0$ and $k_2 = 0$ requires a comparison of the characters of the representations in Tables 4.3 and 4.5. To this end, in both tables consider the characters of the classes E and $\sigma(xz)$ in C_{2v} and E and σ in C_{1h}.

In each of the groups the cited characters correspond to the group operations E and $R_2(2)$. For a comparison, the relevant characters have been taken from the irreducible representations A and B in Table 4.3 and listed below in Table 4.6. In addition, the classes E and $\sigma(xz)$ corresponding to the operations E and $R_2(2)$ in Table 4.5 have been listed in Table 4.6 and labeled by the irreducible representations they are from.

Table 4.6: Table of the E and σ characters of the group elements E and R_2 (2)

	E	$\sigma(xz) = \sigma$
A	1	1
B	1	-1
A_1	1	1
A_2	1	-1
B_1	1	1
B_2	1	-1

From a comparison of the entries in Table 4.6 it is seen that upon leaving the point X at $k_1 = \frac{\pi}{a}$ and $k_2 = 0$ and moving along the Δ line ($k_1 \neq 0$ and $k_2 = 0$) requires the representations A_1 and B_1 to move to A. Similarly, the representations A_2 and B_2 move to B.

An Example

As an example of the application of the ideas of symmetry in the determination of the modal degeneracies, consider a recent study by Hergert et al. [34]. This presented a treatment of the band structure of a square lattice photonic crystal and the resolution of its modes into representations of the point group symmetries of the square lattice. In particular, the considerations are made of a square lattice photonic crystal composed of circular cylindrical air columns periodically embedded in a dielectric medium with $\varepsilon = 2.1$ and a filling factor of $f = 0.5$. Both the TM and TE results are treated, with the lines and points in the band structure labeled by the irreducible representations discussed earlier.

A summary of the results of the symmetry considerations of the modal degeneracies is presented in Table 4.7a and b, and these tables are discussed in the remainder of this section. The emphasis here is that the group theory considerations provide information about the possible modal degeneracies of the system, even in the absence of the computed dispersion relation. The group theory considerations do not, however, fix the energy of the modes or the ordering in energy of the different irreducible representations present in the system. These are determined by the details of the dielectric constants and filling fractions of the particular systems considered.

The symmetry points of the square lattice Brillouin Zone are labeled as in Fig. 4.5. A point in the Brillouin Zone with the highest symmetry occurs at Γ (i.e., at $k_1 = k_2 = 0$). At this point the appropriate square lattice symmetries correspond to those of the group C_{4v} given in Table 4.1, with their corresponding irreducible representations given in Table 4.2. From the irreducible representations of the C_{4v} group presented in Table 4.2 it is seen that the modes at Γ must be composed of double degenerate modes of the two-dimensional representation denoted by E and singly degenerate modes from the four different one-dimensional representations denoted as $A_1, A_2, B_1,$ and B_2. Consequently, the eigenmodes at this point are either

Table 4.7: (a) Representations at symmetry points of the Brillouin Zone and (b) representations at symmetry lines of the Brillouin Zone

Table 7a

Points in BZ	TM Pt. Group	TM	TE Pt. Group	TE
M	C_{4v}	B_2, E, A_1	C_{4v}	A_1, B_1, E
Γ	C_{4v}	A_1, B_1, E, A_1	C_{4v}	A_1, A_1, B_1, E
X	C_{2v}	B_1, A_1, A_2, B_1, B_2	C_{2v}	A_1, B_1, B_1, A_2

Table 7b

Lines in BZ	TM Pt. Group	TM	TE Pt. Group	TE
Σ	C_{1h}	A, B, A	C_{1h}	A, A, B
Δ	C_{1h}	A, A, B, A	C_{1h}	A, A, A, B

singly or doubly degenerate. Results from the study in Ref. [34] for both the TM and TE polarized modes are presented in Table 4.7a and found to be in general agreement with the modal degeneracies of C_{4v} in Table 4.2.

The symmetries of the lowest energy modes for the square lattice treated in Ref. [34] are listed in Table 4.7. Each irreducible representation is presented, ordered with its modal energy increasing from left to right, starting from the lowest energy mode. In this listing it should be remembered that the modes associated with the E representation are two degenerate modes while the modes of the other representations are singly degenerate. In both the TM and TE modes the lowest energy mode is seen to be singly degenerate so that the doubly degenerate modes occur at higher energies. However, it is important to note that the energy ordering of the modes may differ in other realizations of the square lattice symmetry.

Similar to the Γ point, the point labeled by M (i.e., for $k_1 = k_2 = \frac{\pi}{a}$) is a high symmetry point corresponding to the C_{4v} point group. In Table 4.7a the lowest energy modes of M are again listed with the modal energy increasing from left to right starting from the lowest energy mode. Again, for the particular system studied in Ref. [34], the lowest energy modes are singly degenerate and the doubly degenerate E modes occur at higher modal energies. This energy ordering, however, may differ from that in other square lattice photonic crystal systems.

A difference in the modal degeneracies is encountered at the X symmetry point. At X the modal wavevector is $(k_1, k_2) = (\frac{\pi}{a}, 0)$ and the symmetry properties of the system are defined by the C_{2v} point group symmetry shown in Table 4.5. In the C_{2v} group only one-dimensional irreducible representations exist so that the modal energies are all singly degenerate.

The earlier discussions are of the symmetries of particular points of symmetry of the square lattice. There are also symmetries associated with lines in the Brillouin Zone of the square lattice. Specifically, Σ represents a point along the line between Γ and M and Δ represents a point along the line between Γ and X. The symmetry group at the points on these lines is C_{1h} which is shown in Table 4.3. All of the irreducible representations of the C_{1h} group are one-dimensional, and, consequently, there are only singly degenerate modes along these lines. The one-dimensional representation of C_{1h} are denoted as A and B, and their energy ordering for the lowest energy modes of the study in Ref. [34] are listed in Table 4.7b.

In the next section, the discussions turn from the application of group theory to determine the possible modal degeneracies of a photonic crystal from its geometric properties to the evaluation of the energies of the modal solutions.

4.6   PHOTONIC AND PHONONIC CRYSTALS AND THEIR APPLICATIONS

The band structures of photonic [1, 2, 4, 11, 21, 22] and phononic crystals [7, 20, 23] have been studied using a variety of methods. These include planewave expansion techniques, computer simulation methods, and determinantal techniques. All of these methods allow for the highly accurate evaluation of the dispersion relations and wavefunctions of modes of the system in a quick, efficient manner.

In the planewave method the Helmholtz wave equation [11, 32] in the periodic medium is Fourier transformed to be expressed in wavevector space. This yields a system of linear equations for the Fourier coefficients of the modal wavefunctions in wavevector space, with the system of equations presented in the form of an eigenvalue problem for the modal frequencies in terms of the modal wavevectors. In this formulation the dielectric properties of the system enter the eigenvalue equations as the Fourier transforms of the periodic dielectric functions of the media.

Alternatively, computer simulation techniques for the determination of the modal wavefunctions and their frequencies are generally based on two commonly applied approaches [7, 11, 20–23]. In one approach the Maxwell equations (Newton equations) in the time domain are directly integrated numerically as a set of first order partial differential equations of the photonic (phononic) crystal media [11, 21]. This methodology has found many applications in the determination of band structures and the treatment of waveguides and impurity modes in photonic and phononic crystals [7, 11, 20–23].

A second simulation method involves the study of the Helmholtz wave equations of the optical or acoustic modes in the frequency domain [11]. This method approaches the solutions of the Helmholtz equation variationally or in terms of a Green's function treatment such as in the method of moments [11].

The determinantal method [11] is based on discretizing the Helmholtz problem into a linear set of algebraic equations which take the form of an eigenvalue problem. The modal frequencies of the algebraic system are obtained in a frequency search for those frequencies which set the determinate of the homogeneous eigenvalue equations to zero. This last method is particularly effective in problems in which the dielectric or acoustic properties of the system being studies are themselves frequency dependent [11].

A variety of physical systems have been treated by these methods and some of them will now be discussed. The focus will be on two-dimensional systems for both photonic and phononic cyrstals [11, 21]. These are types of systems which have found applications in basic optical and acoustic circuit applications, particularly in optoelectronics. First, some photonic crystal results will be presented. This is followed by some examples of phononic crystals.

4.6.1   PHOTONIC CRYSTAL SYSTEMS

The earliest studied systems involved nonconducting dielectric structures [11, 21]. Common geometries treated for these systems included periodic patterns of dielectric cylinders in vacuum or of vacuum cylinders in a dielectric background. The most common patterns investigated involved the square lattice and the triangle lattice. At one point it was thought that only the triangle lattice could exhibit both TM and TE modes with a common band of stop band frequencies. It is, however, now known that both square and triangle system exist which have a common set of overlaping stop band frequencies.

In Fig. 4.6 an example of a band structure of a two-dimensional photonic crystal on a square lattice is presented [35]. This is a system of dielectric cylinders of circular cross section and dielectric constant $\varepsilon = 9$ with a background media of air. The filling fraction of the cylinders is $f = 0.4488$ and the plot of frequency, ω, vs. wavevector is made along a number of different directions of high symmetry in the wavevector space represented in Fig. 4.5.

The results in Fig. 4.6 were computed [35] using the planewave technique for a system of approximately 400 planewaves. As a comparison, the system in Fig. 4.6 was also studied experimental [36] for an experimental realization formulated to exhibit a band structure at microwave frequencies. The experimental data measured for the system are presented in Ref. [35], and the agreement between theory in Fig. 4.6 and experiment in Figs. 1 and 3 of Ref. [36] is found to be very good.

The results in Fig. 4.6 involve a model with a frequency-independent dielectric. These are easiest handled by the method of planewave expansion. More complex methods, however, are required in the study of frequency-dependent dielectric features. Methods for the treatment of these problems rely on the conditions of solvability of a system of linear algebraic equations. By searching for the frequencies which allow for solutions of the eigenvalue equations representing the system, the dispersion relation is mapped out.

As an example of the application of the determinantal method to handle a system with a frequency-dependent dielectric constant, consider a model of metallic cylinders embedded in vacuum [24]. In Fig. 4.7 some results are presented of a study of metal cylinders in a vacuum background. The dielectric constant of the metal in the cylinders is of the form $\varepsilon(\omega) = 1 - \frac{\omega_p^2}{\omega^2}$ where ω_p is the plasma frequency of the metal, and the filling fraction of the circular cross-section cylinders is 0.1%.

The frequency dependence of the metal cylinders introduces a new aspect to the band structure of the square lattice photonic crystals. In general, because of the small filling fraction

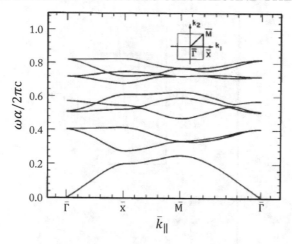

Figure 4.6: Band structure of a square lattice photonic crystal of dielectric cylinder with dielectric constant $\varepsilon = 9$ in an air background. The filling fraction of the cylinders is $f = 0.4488$ and the nearest neighbor separation between the cylinders is a. The inset indicates the region of $k-$space containing a complete set of unique modal solutions [35]. Reprinted with permission from Ref. [35], Optical Society of America.

of the model the dispersion relation of the photonic crystal is little changed from that of free space. However, as an exception to this a new series of flat bands is introduced into the system at low frequencies. These occur near the zero of the metallic dielectric constant. This is new feature in the photonic crystal band structure arising from the dispersive nature of the dielectric properties of the metal.

The interesting properties in the band structure of the two different system in Figs. 4.6 and 4.7 are the stop and pass bands in Fig. 4.6 and the series of flat bands in Fig. 4.7. Both of these types of features have entered into a variety of new and important technological applications in optoelectronics [11].

4.6.2   PHONONIC CRYSTAL SYSTEMS

Next, some results for the band structure of two-dimensional phononic crystals [7, 20, 23] are present as illustrations of typical properties of these systems and for a comparison with photonic crystal band structures. For these discussions the focus is on square lattice systems as the earlier photonic crystals and discussions of symmetry properties were made for these lattice types.

In the study of phononic crystals, crystals formed as arrangements of solid-solid, solid-liquid, and liquid-liquid components have all been treated [7, 20, 23]. Each of these types of systems exhibit a variety of polarization and dispersive properties in their modal structures. As

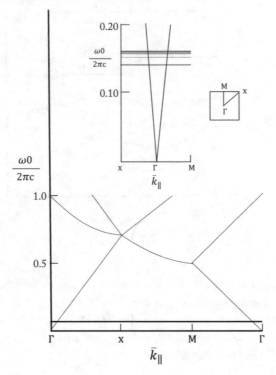

Figure 4.7: Band structure of a square lattice of metal cylinders for the case in which $\frac{\omega_p a}{c} = 1$ [24]. Reprinted figure with permission from Ref. [24]. Copyright 1993, America Physical Society.

shall be seen later in the discussion of metamaterials, solid-liquid mixtures can be particularly important for the development of new features found in some of the recently engineered materials. These discussions in the metamaterial [7] context will be held for later, and in the following only a solid-solid band structure example is presented. The reader is referred to the literature for the full variety of band structure examples that have been discussed for solid-liquid and liquid-liquid systems [7].

In Fig. 4.8 results are presented for the band structure of a square lattice phononic crystal [20]. The system is formed of an array of infinitely long quartz cylinders that are surrounded in an epoxy background. For the calculations numerical simulations based on finite element methods were used, and in the original reference [20] a study was presented comparing the results for the infinite cylinders with results for a slab composed as an array of finite length cylinders formed on the same square lattice. The filling fraction of the cylinders which were of a circular cross section was $f = 0.5$. In the determination of the band structure all possible modal polarizations where included [20].

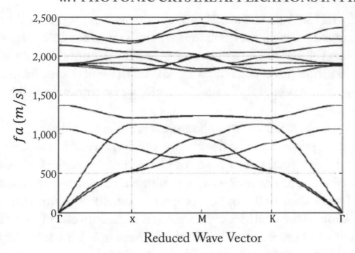

Figure 4.8: Band structure of a square lattice phononic crystal composed of quartz cylinders embedded in an epoxy background. Reprinted figure with permission from [20]. Copyright 2006, American Physical Society.

The dispersion relations observed in the phononic system, generally, are similar to those found in the photonic crystals. In particular, there are a series of stop and pass bands in both the phononic and photonic crystals. As with photonic crystals, the stop bands of the phononic crystal are available to suppress propagation in the same band of frequencies along every direction for motion in the plane perpendicular to the cylinder axes. This allows for the design of phononic waveguides and circuits.

In addition, the modified photonic and phononic dispersion relations can also enhance or diminish the modal density of states in frequency over those of free space. This results in the enhancement or suppression of atomic or vibrational transitions which are dependent on the number of modes available into which excitation processes can decay [11]. In this regard, these manipulations of the density of modes in space allow for the development of media with important new physical properties.

Another important application of photonic crystal design is in the enhancement of the properties of traditional fiber optics. Some discussion of these are now given.

4.7   PHOTONIC CRYSTAL APPLICATIONS IN FIBER OPTICS

A brief introduction to some applications of photonic crystals in fiber optics is now presented [37]. The idea is to develop some basics of fiber optics technology and to understand how concepts from photonic crystal technology can be used to enhance the performance of optical fibers. The general idea behind the design of optical fibers is to create a fiber strand of optical materials which transmits optical energy along its length while confining the energy within the proximity of the axis of the fiber. The confining mechanism in optical fibers arises from the mismatch of the optical properties of the fiber material and those of the media exterior to the fiber. In this regard, in optics, the optical fiber is the analogy of a copper or silver wire in electronic circuit technology.

Photonic crystal fibers incorporate photonic crystals in their design. The photonic crystal is introduced into the optical media of the fiber as a periodic variation of the optical properties in the plane perpendicular to the axis of the fiber. Consequently, the fiber remains translationally invariant along its axis. The photonic crystal patterning may be a two-dimensional lattice pattern or may be a periodic variation of the optical properties radially outwards from the fiber axis.

The introduction of the photonic crystal pattern is made in one of two different ways [11, 37]. In the first type of photonic crystal fiber, the patterning is used to affect a gradient of the dielectric properties of the optical medium. This is the limit in which the medium responds as an effective medium, and the stop band properties of the periodicity of the media do not directly enter the fiber operation. The slow variation of the dielectric properties in the radial direction is used to enhance the confinement properties of the fiber over those of conventionally clad fibers [37].

In the second type of system the stop band structure of the periodic medium opens frequency stop bands which help to confine the radiation within the fiber. In these types of systems, the confinement properties of the fiber either arise from or are enhanced by the presence of the stop band structure.

In the following, an elementary treatment of the equations for a cylindrical fiber of circular cross section is given based on a highly simplified model. This is meant to give the general idea of what is involved in the design of optical fibers. Following this some brief discussions are given of the closely related problem of optical resonators and their applications along with optical fibers in the design of lasers.

4.7.1 MODEL OF AN INFINITE STRAIGHT WAVEGUIDE OF CIRCULAR CROSS SECTION

As an illustration of the fiber optic waveguide, consider the simple model of a straight cylindrical dielectric waveguide which is of infinite length [30]. A schematic of such a waveguide is shown in Fig. 4.9. The fiber has a circular cross section in the x_1–x_2 plane and the fiber axis is along the x_3-space axis. The dielectric constant of the fiber $\varepsilon(x_1, x_2)$ only depends on x_1 and x_2 but is independent of x_3. In this regard, optical fibers are often designed to have a radial variation of their dielectric constant. As shall be discussed later, this radial variation is put in the fiber as an aid in the confinement of traveling electromagnetic waves bound to the fiber.

Outside of the cylindrical fiber the region is often filled by air, and the design of the fiber is made to support electromagnetic solutions in which the fields outside the fiber decay with

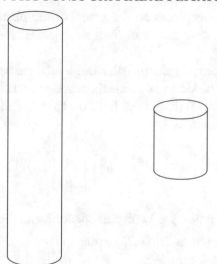

(a)   (b)

Figure 4.9: Schematics of: (a) an infinitely long cylinder waveguide of radius R and (b) a cavity resonator of radius R and height d. For both (a) and (b) the axes of cylindrical features are the x_3 spatial axis.

separation from the fiber axis. The solutions inside and outside of the fibers are matched to one another by a set of electromagnetic boundary conditions.

Inside and outside the fiber waveguide the fields are described by the Maxwell equations in the absence of charge and current sources. These are of the general form [30]

$$\nabla \times \vec{E} = i\frac{\omega}{c}\vec{B}, \quad \nabla \cdot \varepsilon\vec{E} = 0 \tag{4.73a}$$

$$\nabla \times \vec{B} = -i\frac{\varepsilon\omega}{c}\vec{E}, \quad \nabla \cdot \vec{B} = 0, \tag{4.73b}$$

where ε is the appropriate dielectric constant for the different regions considered in the waveguide problem.

Due to the translational symmetry of the optical fiber along the x_3 direction, the field solutions have the general form of planewaves bound to and propagating along the fiber axis. An expression for such solutions is

$$\vec{E}(x_1, x_2, x_3) = \vec{E}(x_1, x_2)\,e^{\pm ikx_3 - i\omega t} \tag{4.74a}$$

$$\vec{B}(x_1, x_2, x_3) = \vec{B}(x_1, x_2)\,e^{\pm ikx_3 - i\omega t}. \tag{4.74b}$$

These field solutions place an emphasis on the nature of the planewave propagation in the x_3 direction along the translationally invariant fiber as separate from the behavior of the fields transverse to the fiber axis.

As an example of the forms in Eq. (4.74), consider the particularly simple case in which ε is position independent both inside and outside the waveguide. It follows from substituting the forms in Eq. (4.74) into Eq. (4.73) that in this limit the electric and magnetic fields are solutions of the Helmholtz equations given by [30]

$$\left[\nabla_t^2 + \varepsilon \frac{\omega^2}{c^2} - k^2 \right] \left\{ \begin{array}{c} \vec{E} \\ \vec{B} \end{array} \right\} = 0. \tag{4.75}$$

Here, $\nabla_t^2 = \frac{\partial^2}{\partial x_1^2} + \frac{\partial^2}{\partial x_2^2}$ is the transverse Laplacian operator, and the dependence of the solutions in the transverse plane are seen from Eq. (4.75) to depend both on the frequency and wavevector components of the modes along the x_3-axis.

The field solutions inside and outside the fiber are matched by the electromagnetic boundary conditions, and this matching sets the fields and the field dispersion relations of the modal solutions. The solutions described are those of a basic model of an optical fiber which, while interesting, are too simplistic for calculations of systems applied in technology. A more general formulation will now be presented which offers an approach to practical results.

For realistic waveguides applied in technology, a treatment of optical fibers which includes the case in which $\varepsilon(x_1, x_2)$ depends on the coordinates x_1 and x_2 is needed [30]. The approach for these types of systems is more complicated than the optical fiber guide composed of a uniform homogeneous dielectric and may require simulation methods for their treatment. The solutions for the position-dependent dielectrics in general have a much more complicated structure of modes and their dispersion relations than the simpler systems mentioned earlier.

It is useful in the following discussions of the solutions of Eqs. (4.73) and (4.74) for $\varepsilon(x_1, x_2)$ dependent fibers to separate the vector Maxwell equations in Eq. (4.73) into sets of equations projected along the x_3 direction and those projected onto the x_1–x_2 plane. In this separation, it is useful to introduce the notations $\nabla_t = \hat{x}_1 \frac{\partial}{\partial x_1} + \hat{x}_2 \frac{\partial}{\partial x_2}$, $\vec{E}_t = \hat{x}_1 E_1 + \hat{x}_2 E_2$, and $\vec{B}_t = \hat{x}_1 B_1 + \hat{x}_2 B_2$ applied to the fields and operator components in the x_1–x_2 plane, and the notations $\vec{E}_3 = E_3 \hat{x}_3$ and $\vec{B}_3 = B_3 \hat{x}_3$ in the x_3 direction .

Applying these notations, Faraday's law becomes [30]

$$\hat{x}_3 \cdot \left(\nabla_t \times \vec{E}_t \right) = i \frac{\omega}{C} B_3 \tag{4.76a}$$

for the projection along the x_3 direction, and the components in the x_1–x_2 plane are given by

$$\frac{\partial \vec{E}_t}{\partial x_3} + i \frac{\omega}{c} \hat{x}_3 \times \vec{B}_t = \nabla_t E_3. \tag{4.76b}$$

A similar application to Ampere's law gives a projection along the x_3 direction of the form

$$\hat{x}_3 \cdot \left(\nabla_t \times \vec{B}_t \right) = -i\varepsilon \frac{\omega}{C} E_3 \tag{4.77a}$$

and the components in the x_1–x_2 plane become [30]

$$\frac{\partial \vec{B}_t}{\partial x_3} - i\varepsilon \frac{\omega}{c} \hat{x}_3 \times \vec{E}_t = \nabla_t B_3. \tag{4.77b}$$

From the sets of equations, Eqs. (4.73), 4.76, and (4.77), a set of equations can be generated which are to be solved for the field components in the x_3 direction. In addition, it is shown from Eqs. (4.76) and (4.77) that the fields in the transverse plane can be written in terms of the field solutions in the x_3 direction. In the following, first the transverse fields will be written in terms of the solutions for the fields in the x_3 direction. This is followed by a discussion of the equations for the field components in the x_3 direction.

Using the relations $\frac{\partial \vec{E}_t}{\partial x_3} = \pm ik\vec{E}_t$ and $\frac{\partial \vec{B}_t}{\partial x_3} = \pm ik\vec{B}_t$ in Eqs. (4.76b) and (4.77b) transforms them into the set of equations [30]

$$\pm ik\vec{E}_t + i\frac{\omega}{c}\hat{x}_3 \times \vec{B}_t = \nabla_t E_3, \tag{4.78a}$$

and

$$\pm ik\vec{B}_t - i\varepsilon\frac{\omega}{c}\hat{x}_3 \times \vec{E}_t = \nabla_t B_3. \tag{4.78b}$$

In a next step applying $\hat{x}_3\times$ to Eq. (4.78b) transforms this equation to the form

$$i\varepsilon\frac{\omega}{c}\vec{E}_t \pm ik\hat{x}_3 \times \vec{B}_t = \hat{x}_3 \times \nabla_t B_3. \tag{4.79}$$

The resulting Eqs. (4.78a) and (4.79) are now a set of equations which can be solved for $\vec{E}_t$ and $\hat{x}_3 \times \vec{B}_t$. From these solutions the field $\vec{B}_t$ is obtained by applying the identity $\hat{x}_3 \times \left(\hat{x}_3 \times \vec{B}_t \right) = -\vec{B}_t$.

In this way it follows that at the end of the process [30]

$$\vec{E}_t = \frac{i}{\gamma^2} \left[\pm k\nabla_t E_3 - \frac{\omega}{c}\hat{x}_3 \times \nabla_t B_3 \right] \tag{4.80a}$$

and

$$\vec{B}_t = \frac{i}{\gamma^2} \left[\varepsilon\frac{\omega}{c}\hat{x}_3 \times \nabla_t E_3 \pm k\nabla_t B_3 \right] \tag{4.80b}$$

where $\gamma^2 = \varepsilon\frac{\omega^2}{c^2} - k^2$. From the form of the equations generated, it is seen that Eqs. (4.80) allows for the transverse fields to be obtained directly from the solutions for the fields in the axial direction of the fiber.

Once a solution is obtained for $\vec{E}_3$ and $\vec{B}_3$ from the Helmholtz equations for these two field components, the transverse field components are obtained from Eqs. (4.80). To obtain the Helmholtz equations for $\vec{E}_3$ and $\vec{B}_3$ involves a reconsideration of the Maxwell equations in Eqs. (4.73).

Taking the curl of Faraday's law in Eqs. (4.73) and recognizing the position dependence of the dielectric constant of the optical fiber within the x_1–x_2 plane gives the inhomogeneous Helmholtz equations for the electric and magnetic fields. These field equations are written as [30]

$$\nabla^2 \vec{E} + \varepsilon \frac{\omega^2}{c^2} \vec{E} = -\nabla \left(\frac{1}{\varepsilon} \vec{E} \cdot \nabla \varepsilon \right) \tag{4.81a}$$

and

$$\nabla^2 \vec{B} + \varepsilon \frac{\omega^2}{c^2} \vec{B} = i \frac{\omega}{c} \nabla \varepsilon \times \vec{E}. \tag{4.81b}$$

It is seen from both of the equations in Eqs. (4.81) that the electric field and the magnetic induction are coupled to one another by the position dependence of the dielectric constant. This was not the case in Eq. (4.75) for the position-independent fiber dielectric constant and results in a much more complicated field structure in the optical fiber with position-dependent dielectric properties.

Solutions for the x_3 components of the electric field and magnetic induction are now generated from Eqs. (4.81). Taking the dot product of the left- and right-hand sides of Eqs. (4.81) followed by a little algebra gives [30]

$$\nabla_t^2 E_3 + \gamma^2 E_3 = \pm \frac{1}{\varepsilon} \frac{\omega}{c} \frac{k}{\gamma^2} \hat{x}_3 \cdot (\nabla_t \varepsilon \times \nabla_t B_3) + \frac{1}{\varepsilon} \frac{k^2}{\gamma^2} \nabla_t \varepsilon \cdot \nabla_t E_3 \tag{4.82a}$$

and

$$\nabla_t^2 B_3 + \gamma^2 B_3 = \pm \frac{-\omega}{c} \frac{k}{\gamma^2} \hat{x}_3 \cdot (\nabla_t \varepsilon \times \nabla_t E_3) + \frac{\omega^2}{c^2 \gamma^2} \nabla_t \varepsilon \cdot \nabla_t B_3. \tag{4.82b}$$

These are a coupled set of partial differential equations for the axial components of the electric and magnetic fields. Notice, however, that the coupling terms between the fields vanish in the absence of a position dependent dielectric. In general, an analytic solution is not possible for these two equations and some numerical techniques are needed for an accurate study of them.

In the treatment of a straight circular fiber surrounded by air, Eqs. (4.82) is solved and matched with electromagnetic boundary conditions at the surface of the cylinder to solutions in the region of air. For this matching, the solutions of interest are bound to the fiber, propagating along the fiber parallel to its axis. From Eqs. (4.80) it then follows that the transverse electric and magnetic fields are obtained from the axial fields. The complete solutions finally consist of a set of different guided modes along with their dispersion relations.

A simplified example will now be given to illustrate some of the basic features found in the general optical fiber solutions.

4.7.2   EXAMPLE OF A CYLINDRICAL WAVEGUIDE WITH PERFECT CONDUCTING WALLS

As a simple example of a waveguide system, meant to give the impression of the general structure found in the solutions of fiber optical waveguide modes, consider the guided modes within a perfect conducting cylindrical shell [30]. The axis of the shell is taken along the x_3-axis and the radius of the shell in the x_1–x_2 plane is R. For simplicity, inside the cylinder is a uniform isotropic homogeneous dielectric with dielectric constant, ε. The medium within the region inside the waveguide channel is now treated as position independent.

Within the bulk of the perfect conducting shell the fields satisfy $\vec{E} = \vec{D} = \vec{B} = \vec{H} = 0$. As a consequence of the absence of fields in the conductor, it follows that at the interface between the dielectric medium within the cylinder and the perfect conducting surface [30]

$$E_3|_S = 0 \tag{4.83a}$$

and

$$\vec{B}_t|_S = 0. \tag{4.83b}$$

The behaviors of the other two $\vec{D}$ and $\vec{H}$ fields at this interface are involved with the existence of surface charges and surface currents in the system and are not of interest in the following.

In the presentation that follows, Eq. (4.83a) is a first boundary condition that is employed in setting the solutions, and a second boundary condition is obtained when Eq. (4.83b) is applied to the form of Ampere's law projected into the x_1–x_2 plane. In this manner, taking the dot product of Eq. (4.77b) with the normal to the perfect conducting surface, $\hat{n}$, it follows from Eq. (4.83b) that at the surface [30]

$$\hat{n} \cdot \nabla_t B_3 = \frac{\partial B_3}{\partial n}|_S = 0. \tag{4.83c}$$

At the perfect conducting surfaces the resulting conditions in Eqs. (4.83a) and (4.83c) are used to determine the field components in the axial direction of the various guided modes of the system. The focus in the following is on obtaining the solutions of the axial fields of the guided modes and using these solutions to obtain the transverse fields of the modes from the Maxwell equations.

The study of the modal solutions begins by solving for the x_3 components of the electric and magnetic fields. These are obtained from the Helmholtz equations

$$\left[\nabla_t^2 + \varepsilon \frac{\omega^2}{c^2} - k^2 \right] \left\{ \begin{array}{c} E_3 \\ B_3 \end{array} \right\} = 0 \tag{4.84}$$

solved subject to the boundary conditions in Eqs. (4.83a) and (4.83c). Once the axial fields are obtained the transverse fields are generated from an application of Eqs. (4.80). There are generally two possible modes in the perfect conducting cylinder waveguide model. These are the so-called sets of TM and TE guided modes and their solutions are now addressed.

One set of solutions of interest are the TM waves. These waves are formed from the condition that $B_3 = 0$, and they have an E_3 which is a solution of Eq. (4.84) subject to the boundary conditions in Eq. (4.83a). The transverse fields of these modes are determined from Eqs. (4.80) which for $B_3 = 0$ takes the form [30]

$$\vec{E}_t = \frac{i}{\gamma^2} \left[\pm k \nabla_t E_3 \right] \tag{4.85a}$$

and

$$\vec{B}_t = \frac{i}{\gamma^2} \left[\varepsilon \frac{\omega}{c} \hat{x}_3 \times \nabla_t E_3 \right]. \tag{4.85b}$$

From Eqs. (4.85) it then follows that the transverse electric and magnetic fields in the TM modes are related by

$$\vec{B}_t = \pm \frac{1}{\left(\frac{kc}{\epsilon \omega} \right)} \hat{x}_3 \times \vec{E}_t. \tag{4.86}$$

A second set of solutions of interest are TE waves for which $E_3 = 0$ while B_3 is a solution of Eq. (4.84) subject to the boundary condition in Eq. (4.83c). Once the axial magnetic field has been found, the transverse fields are determined in terms of B_3 from Eqs. (4.80). In this regard, for $E_3 = 0$ Eqs. (4.80) takes the form [30]

$$\vec{E}_t = \frac{i}{\gamma^2} \left[-\frac{\omega}{c} \hat{x}_3 \times \nabla_t B_3 \right] \tag{4.87a}$$

and

$$\vec{B}_t = \frac{i}{\gamma^2} \left[\pm k \nabla_t B_3 \right]. \tag{4.87b}$$

As with the TM modes, the transverse fields of the electric and magnetic fields are related to one another. This relationship is

$$\vec{E}_t = \pm \frac{-\omega}{ck} \hat{x}_3 \times \vec{B}_t. \tag{4.88}$$

The various TM and TE modes obtained in this way represent the complete set of guided modes for the perfect conducting waveguide. Another type of mode which might be thought possible is the so-called TEM mode which consists solely of transverse electric and magnetic fields. This type of mode is present in other waveguide structures but not in the perfect conducting system.

To see that the TEM mode is not present in the perfect conducting guide consider the form of Faraday's law in Eq. (4.76a) taken in conjunction with Gauss' law. Forming the dot product of Eq. (4.76a) with $\hat{x}_3$ gives

$$\nabla_t \times \vec{E}_t = 0 \tag{4.89a}$$

and for $E_3 = 0$ it follows from Gauss' law that

$$\nabla_t \cdot \vec{E}_t = 0. \tag{4.89b}$$

In this way it is found that the perfect conducting cylinder problem for TEM modes reduces to an electrostatic problem of an interior region surrounded by an equipotential surface. Consequently, $\vec{E}_t = 0$ and TEM modes do not exist in the system. The TE and TM solutions of the system are now obtained.

Consider the TE mode solutions of Eq. (4.84) for B_3 subject to the boundary conditions $\frac{\partial B_3}{\partial n}|_S = 0$. Expressing Eq. (4.84) in polar coordinates and reintroducing the notation $\gamma^2 \equiv \varepsilon \frac{\omega^2}{c^2} - k^2$, it is found that the modal solutions inside the cylinder are given by [30]

$$B_{3,mn}(\gamma_{mn}r, \varphi) = B_{3,0} e^{\pm im\varphi} J_m(\gamma_{mn}r), \tag{4.90}$$

where $J_m(x)$ is the mth cylinder Bessel function, $\gamma_{mn}R$ are the zeros of $\frac{d}{dx} J_m(x) = 0$, n labels the zeros of the derivative of the m Bessel function, and $B_{3,0}$ is the modal amplitude. Here the subscript labels on γ in Eq. (4.90) denote the Bessel function of the mode in Eq. (4.90) and the number of its multiple zeros of the Bessel function derivative. In this scheme, for a given γ_{mn} the frequency, ω, and the wavenumber, k, of the corresponding mode in Eq. (4.90) are related to each other by $\gamma_{mn}^2 = \varepsilon \frac{\omega^2}{c^2} - k^2$. The transverse fields corresponding to the axial field in Eq. (4.90) are obtained from Eqs. (4.87).

Similarly, the TM mode solutions of Eq. (4.84) for E_3 subject to the boundary conditions $E_3|_S = 0$ are found to be solutions inside the cylinder of the from [30]

$$E_{3,mn}(\gamma_{mn}r, \varphi) = E_{3,0} e^{\pm im\varphi} J_m(\gamma_{mn}r). \tag{4.91}$$

Here, $J_m(x)$ is the mth cylinder Bessel function, $\gamma_{mn}R$ are the zeros of $J(x) = 0$, n labels the zeros of the m Bessel function, and $E_{3,0}$ is the modal amplitude. The subscript labels on γ in Eq. (4.91) now denote the Bessel function of the mode and the number of its multiple zeros. In this scheme, for a given γ_{mn}, the frequency ω and the wavenumber k of the corresponding mode in Eq. (4.91) are related to each other by $\gamma_{mn}^2 = \varepsilon \frac{\omega^2}{c^2} - k^2$. The transverse fields corresponding to the axial field in Eq. (4.91) are then obtained from Eqs. (4.85).

The solutions for the electromagnetic fields in a perfect conducting waveguide are found to be represented as linear combinations of the individual modal solutions discussed earlier. A similar structure of solutions is seen in more general types of dielectric waveguides surrounded by air. In this regard, the electromagnetic waves are always sums of modal solutions, but due to the field couplings in Eqs. (4.82) in the dielectric waveguides the classification as TM and TE excitations is not possible for all of the various modes solutions. In addition to the detailed features of the modes being changed from those of the perfect conducting guide, new types of modes are generated which are not related to those found in the perfect conducting waveguide.

Related Problem of Perfect Conducting Resonant Cavity

A closely related problem to the perfect conducting waveguide is that of a cylindrical optical cavity resonator [30]. This is of importance in the study of, for example, surface-emitting photonic crystal lasers, optical multiplexing devices, and sensing mechanisms. Due to the close relationship of the resonator solutions to those obtained for the perfect conducting waveguide, however, it is useful to consider the resonator solutions at this point. These discussions anticipate some of the later technological applications of resonators.

A simple model of a resonator is based on that of the perfect conductor waveguide. The resonator model is formulated by considering a finite waveguide segment which is capped off on its top and bottom by perfect conducting planes. This regards the cavity resonator as a type of conducting "coffee can" which encloses a region of dielectric forming the bulk of the resonator.

A schematic of the resonator is shown in Fig. 4.9. The label region of the can is of circular cross section and radius R in the x_1–x_2 plane, with the axis of the resonator along the x_3-space axis. Perfect conducting plates forming the top and bottom of the closed can are located in the $x_3 = d$ and $x_3 = 0$ planes, respectively. The dielectric within the cavity resonator is a homogeneous isotropic medium with a dielectric constant ε which is independent of x_1, x_2, and x_3.

Inside the resonator the fields are described by the Maxwell equations in Eqs. (4.73) for a system in the absence of charge and current sources. The boundary conditions on the solutions are the same as the waveguide problem with the additional conditions that at the top and bottom of the cans [30]

$$E_t|_{Top} = E_t|_{Bottom} = 0 \tag{4.92a}$$

and

$$B_3|_{Top} = B_3|_{Bottom} = 0. \tag{4.92b}$$

Due to the close relationship between the resonator and waveguide geometries the solution for the resonator fields can be constructed as linear combinations of solutions of the perfect conducting waveguide modes. In this regard, since the waveguide solutions already match the boundary conditions on the label of the can, a linear combination of waveguide solutions is composed to match the boundary conditions in Eqs. (4.92) at the top and bottom of the can.

Proceeding in this manner it is found that the TM solutions satisfying the boundary conditions at the top and bottom of the can are obtained from the waveguide solutions in Eq. (4.91).

The TM mode of the cavity resonator generated from Eq. (4.91) has the form [30]

$$E_{3,mnp}(x_1, x_2, x_3) = E_{3,mn}(\gamma_{mn}r, \varphi) \cos\left(\frac{p\pi x_3}{d}\right) = E_{3,0}e^{\pm im\varphi} J_m(\gamma_{mn}r) \cos\left(\frac{p\pi x_3}{d}\right),$$
(4.93a)

where $p = 0, 1, 2, \ldots$, $J_m(x)$ is the mth cylinder Bessel function, $\gamma_{mn}R$ are the zeros of $J_m(x) = 0$, n labels the different zeros of the m Bessel function, and $E_{3,0}$ is the modal amplitude. From the earlier discussions of the waveguide, the solution in Eq. (4.93a) is seen to satisfies the boundary conditions on the label of the can. In addition, applying Eqs. (4.76) and (4.77) to Eq. (4.93a) generates a transverse electric field which is found to be zero at the top and bottom of the can.

Similarly, the TE solutions are found from Eq. (4.90) to be given by [30]

$$B_{3,mnp}(x_1, x_2, x_3) = B_{3,mn}(\gamma_{mn}r, \varphi) \sin\left(\frac{p\pi x_3}{d}\right) = B_{3,0}e^{\pm im\varphi} J_m(\gamma_{mn}r) \sin\left(\frac{p\pi x_3}{d}\right),$$
(4.93b)

where $p = 1, 2, \ldots$, $J_m(x)$ is the mth cylinder Bessel function, $\gamma_{mn}R$ are the zeros of $\frac{d}{dx}J_m(x) = 0$, n labels the zeros of the derivative of the m Bessel function, and $B_{3,0}$ is the modal amplitude. Consequently, the boundary conditions at the top and bottom surfaces are satisfied as well as those on the label. From the result in Eq. (4.93b) the transverse electric and magnetic fields are obtained applying the results in Eqs. (4.76) and (4.77).

Once the values of $\gamma_{mn}R$ are determined from either of the conditions that $J_m(x) = 0$ or $\frac{d}{dx}J_m(x) = 0$, the modes are set from Eqs. (4.93), (4.76), and (4.75). The frequencies of the modes are obtained from $\gamma_{mn}^2 = \varepsilon\frac{\omega^2}{c^2} - \left(\frac{p\pi}{d}\right)^2$, giving a discrete set of resonances for the set of modes of the cylindrical cavity.

4.7.3   APPLICATIONS OF RESONANT CAVITIES AND PHOTONIC CRYSTAL TECHNOLOGIES

In photonic crystal applications resonant cavities are created by forming an enclosed region of space which is surrounded by photonic crystal material [11]. The modes of the cavity are set at frequencies which are located within the stop band of the surrounding photonic crystal materials. This provides the confinement mechanism of the resonator. A source of stop band radiation is then located within the cavity and set to emit radiation at a stop band frequency.

In this arrangement photonic crystals provide more efficient resonator cavities of higher Q and decreased losses than more conventionally designed systems. For example, cavities made from metal will exhibit Joule loss whereas it is often possible to design a photonic crystal confining of lower loss dielectric media. Similar remarks can be made regarding the photonic optical fibers and more conventional optical fiber designs.

As important general examples of such photonic crystal applications, periodic dielectric patterns embedded into slab geometries have been introduced in the designs of surface emitting lasers, and, additionally, photonic optical fibers have also provided a basis for the design of optical fibers cavities for laser designs [11]. In both systems the photonic crystal components of the designs act to enhance the confinement and energy efficiency of the devices formed by their applications over those of systems based on more conventional technologies. They have also provided mechanisms which offer to lower the lasing threshold.

In the surface emitting laser, a heterojunction is placed in the cavity surrounded by a slab media which has been designed with a periodic confining dielectric pattern. The heterojunction is configured to generate radiation at a frequency in the stop band of the slab media. As a result, the pattern inhibits the entry of radiation into the slab. On the other hand, the dielectric mismatches at the slab surfaces act as the partially reflecting mirror for the laser emissions.

In the photonic crystal fiber laser, a photonic crystal fiber is seeded with rare earth dopants. While the finite length of photonic crystal fiber acts as the confining cavity, the rare earth dopants are the pumped source of the lasing radiation. The photonic crystal fiber laser can enhance the lasing properties over more conventionally designed fiber lasers. In addition, they also extend the fiber laser potential to act as sensor devices and optical amplifiers.

CHAPTER 5

Acoustic and Electromagnetic Metamaterials

Metamaterials are composite media including in their design various engineered features which are tailored to provide the material with a desired response to particular external stimuli [7–17, 22, 39–43]. In these designs a basic goal in the development of metamaterials is to engineer them as media which respond to externally applied excitations as homogeneous materials. In this regard, a second goal of particular importance in their development is that by properly choosing the engineered components, composite media formulations are made displaying new or enhanced ranges of responses to external interactions. The enhancements of properties displayed by metamaterials are generally arrived at through the design of features in their composite which exhibit resonant properties. Consequently, resonant responses are fundamental to the metamaterial designs. This is a point of difference they have with ordinary composites formed of non-resonant materials. Designed in this way, some of the metamaterial responses are of a type not found in naturally occurring optical and acoustic media and, consequently, lead to novel technological applications and designs.

The ability of the metamaterial to appear to be homogeneous depends on the wavelength of the stimulus [7–17, 22, 39–43]. At excitation wavelengths of order of the dimensions of the engineered features of the metamaterial, the structure of the engineered features would be resolved by the excitations. For this limit the detailed structure of the medium is important to its response to the stimulus, and the response is then diffractive in nature. Consequently, the metamaterial under these conditions would not appear to be homogeneous. Only for excitations with wavelengths much larger than the engineered features in the metamaterials can the material give a homogeneous isotropic response, averaging over the structures present in the medium. This case of the limit of homogeneous response is the focus in the design of metamaterials and in their applications presented in this chapter.

The importance of wavelength sensitivity is also seen in the condensed matter physics of crystalline media in which electromagnetic waves with wavelengths of order of the interatomic spacing are diffracted. Here the details of the crystal structure are impressed on the pattern of light scattered from the medium. On the other hand, electromagnetic waves with wavelengths much greater than the interatomic spacings exhibit an average response from the medium, characterized by a dielectric constant. In this long wavelength limit the motion of light is in the domain of refractive propagation [11, 25–27, 30].

Similarly, for vibrational excitations with wavelengths of order of the interatomic spacing the acoustic properties of crystals are characterized by the detailed atomic masses and interatomic bounds [7–17, 22, 39–43]. In this limit the dispersive properties of the vibrational modes exhibit a complex series of frequency bands and the modes of the system are of a complicated nature depending on the details of the atomic arrangements in the material. At wavelengths of sound much larger than the interatomic separations, however, the acoustic properties are described in terms of the mass density and coefficients characterizing the macroscopic elastic properties of the materials, e.g., bulk and Young's moduli [7–9]. This last range is the limit of continuum mechanics which is characterized by a few simple parameters representing the average response of the material.

In the following, some of the recent ideas applied to engineered metamaterials in acoustic and optical systems are presented. The materials are formulated by the addition of various types of acoustic and optical subwavelength resonant features into an otherwise homogeneous medium. In this design the resonant nature of the features forming the metamaterial composite are of fundamental importance to the composition of the material. In particular, these resonant features contribute to an enhancement of the properties exhibited by the resulting metamaterial. The enhancements observed occur in the vicinity of the resonant frequency of the composite features as it is within these frequency regions that the resonators display a wide range of rapidly varying responses from which to select.

In the development of the theory of metamaterials presented in the text the focus is on basic concepts and ideas introduced in terms of simple analytical models. The object will be to focus on the simplest systems which illustrate how metamaterials function without cataloging results from numerous complex computer simulation studies that have been made in the field. For the details of complex systems which are the focus of simulation studies, the reader is referred to the literature for considerations of systems directed to practical applications [7, 9, 11, 22, 25–27].

First, some general remarks are made on the general properties of composite media [39–43]. The virtual crystal and effective media treatment of composite materials are explained. These are the most basic theoretical treatments involving a representation of the composite response to excitations in terms of averages over the physical responses of each of the composites forming the system.

Next, a treatment of acoustic metamaterials is given. This is developed in studies of one-dimensional vibrational chain models into which resonant features are introduced [7–17, 22, 39–43]. The systems studied are shown to exhibit negative effective masses, negative effective spring constants, or both, dependent on the nature of the resonators included in the chain structure. Generalization of these ideas to higher dimensional vibrational systems is then made. The importance of so-called double negative systems which display both negative mass and spring constants are discussed along with the important new physics exhibited by such systems.

In parallel with acoustic metamaterials the ideas of optical metamaterials are presented. The general properties of these systems are based on the properties of split ring resonators introduced into an optical medium [11, 16, 25, 26]. Similar to the acoustic systems, the resonance properties of the split ring resonators are at the basis of the novel properties developed in optical metamaterials. Discussions on the new physics arising from the development and application of optical metamaterials are presented.

The chapter concludes with some discussions of applications of these metamaterials in science and technology. These applications are found both in the design of bulk and surface metamaterial components and include the important new technology of media exhibiting negative refractive index.

5.1   EFFECTIVE MEDIA METHODS FOR COMPOSITE MEDIA

When treating the response of composite media to external stimuli it is often useful to view the composite as being approximated by a homogeneous isotropic effective medium [11, 40, 42]. This, in particular, is the case when the wavelength of the external stimuli is large compared to the size of the grains or spatial features comprising the composite medium. In this point of view the response of the medium is obtained as an average over a region which is small compared to the wavelength but large compared to the individual grains forming the system.

For this first section of the chapter some of the basic types of theoretical approaches that have been applied to study composites as represented in terms of an effective medium are outlined. These include the virtual crystal and coherent potential types of approaches, each of which offers a first rough approximation for dealing with the physics of such materials. They provide insight into the origins of the response of the system which are often not obvious from computer simulation studies of the detailed models developed for composite media.

Virtual Crystal Method

The simplest treatment to approximate the response of a composite medium is the virtual crystal method. This is based on the simple idea of direct averaging. In this regard, consider a composite formed of mesoscopic grains of different media. For example, the medium could be formed of dielectric grains of different permittivities, $\{\varepsilon_i\}$, though any macroscopic property could be similarly treated for the granular mixture.

If grains of permittivities in the set $\{\varepsilon_i\}$ occur in the system with corresponding volume fractions $\{f_i\}$, then a reasonable crude approximation of the homogeneous response of the system involves a simple average. The average response of a material of N dielectrics $\{\varepsilon_i\}$ would be described by the effective permittivity given by [39, 41]

$$\varepsilon_{eff} = \sum_{i=1}^{N} f_i \varepsilon_i, \qquad (5.1)$$

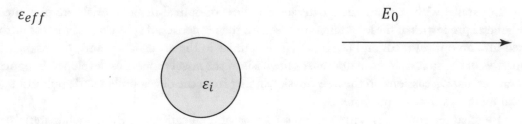

Figure 5.1: A dielectric sphere of radius a and dielectric constant ε_i centered at the origin of coordinates. Outside is an effective medium with dielectric constant ε_{eff} and the average uniform field E_0 directed along the z-direction in the effective medium.

where $\sum_{i=1}^{N} f_i = 1$. The versatility of this approach to treat other properties is evident. For example, the generalization of Eq. (5.1) to the speed of sound $\{c_i\}$ of the acoustic response in the same medium yields an effective speed in the composite given by $c_{eff} = \sum_{i=1}^{N} f_i c_i$.

Coherent Potential Method

A more sophisticated treatment is an effective medium approach known as the coherent potential method. In this approach each of the grains of the composite are treated as interacting with an effective medium representing the average properties of the rest of the grains in the system.

To see how the coherent potential approach works [11, 39, 41], consider a set of spherical grains of radius a with N different dielectric constants $\{\varepsilon_i\}$ and volume fractions $\{f_i\}$. These represent the grains in a composite medium and each of these spheres is taken to interact with the effective medium which is to be chosen to represent the response of the composite medium. The interaction of the spherical grain of dielectric ε_i with the effective medium is shown in Fig. 5.1.

If a uniform electric field E_0 is applied in the homogeneous effective medium along the z-axis in Fig. 5.1, the electric potential throughout all space can be solved as a boundary value problem of the Laplace equations. This is treated in all texts on electrodynamics and the results can be used here to express the electric potential throughout all space in terms of the parameters in Fig. 5.1.

From these treatments it is found that in terms of the situation in Fig. 5.1 the electric potential is [30]

$$\phi_{in}^i = -\frac{3}{2 + \frac{\varepsilon_i}{\varepsilon_{eff}}} E_0 r \cos \theta \tag{5.2a}$$

inside the sphere, and

$$\phi^i_{outside} = -E_0 r \cos\theta + \frac{\dfrac{\varepsilon_i}{\varepsilon_{eff}} - 1}{\dfrac{\varepsilon_i}{\varepsilon_{eff}} + 2} \frac{a^3 E_0}{r^2} \cos\theta \qquad (5.2b)$$

outside the sphere. Performing an average of the results in Eqs. (5.2) over the N different ε_i in the system should reproduce in a self-consistent manner the potential of the uniform field E_0 existing in the effective medium.

In the effective medium approximation ε_{eff} is then chosen such that [39, 41]

$$\sum_{i=1}^{N} f_i \phi^i_{outside} = -E_0 r \cos\theta. \qquad (5.3)$$

This requires that the field outside the sphere on average is only the applied field E_0, i.e., the dipolar contributions outside the sphere average to zero. A consequence is that the effective dielectric is obtained as a solution of

$$\sum_{i=1}^{N} f_i \frac{\varepsilon_i - \varepsilon_{eff}}{\varepsilon_i + 2\varepsilon_{eff}} = 0 \qquad (5.4)$$

where $\sum_{i=1}^{N} f_i = 1$. In this way, the ideas leading to Eq. (5.4) form the basis of an effective medium approximation which is straightforwardly extended to treat many other properties of the system, e.g., the speed of sound.

An importance of the virtual crystal and effective medium approximations is that they show how a variety of materials with designer properties can be formulated in terms of composite materials by incorporating the different properties of naturally occurring media. An indication of the range of possible response available from composite technologies is also provided in this way. Both of these indicators are useful factors in guiding the design of systems for technological applications.

As shall be seen later in this chapter, it is also possible to incorporate a variety of engineered nano-features into materials that form part of the components used to make composites. Many of these engineered composites can be made to exhibit responses as homogeneous media which are not otherwise found in nature. These abilities are the basis of the so-called metamaterials which are presently discussed.

Particularly important in the design of engineered metamaterials are engineered nano-features which exhibit resonant properties as a function of frequency. Such resonant features may allow for engineered materials that exhibit, for example, frequency regions of negative mass, negative Young's modulus, or negative permeability which are not found in non-engineered, naturally occurring, materials. This opens up new possibilities in design technologies.

Some examples of the importance of resonant features in the design of acoustic and electromagnetic metamaterials are now discussed. The presentation focuses on developing these properties using simple models which provide the basic ideas of metamaterial designs. More detailed models applied in device technology often require computer simulation studies to tailor the material response to the design requirements of the system being built.

5.2   RESONATOR BASED METAMATERIALS

In this section some basic models are studied which allow for the introduction of resonant features into dynamical systems [7–17, 22, 39–43]. The resonant features are important as they contribute to and sometimes radically alter the frequency response of the media into which they are embedded. In particular, the resonator response to a driving frequency varies rapidly as an applied frequency is tuned through the resonance of the resonator units. By properly choosing the characteristics of the resonator, metamaterials containing them can be made to exhibit unusual features not found in naturally occurring materials. The resonance properties of the metamaterials for wavelengths at which they can be regarded as homogeneous media, then opens new possibilities in the study of acoustic and optical systems. Metamaterials formed in this way are a type of engineered composite media which can be tailored to meet applications.

In order to provide simple analytic treatments a focus in the following is on the properties of one-dimensional media [10, 12–15, 17, 19, 41, 42]. One-dimensional models are exactly solvable and provide a clear demonstration of the origins of physical properties that are often obscured in systems which require computer simulation studies. Many of the ideas introduced in these one-dimensional systems can, however, be straightforwardly extended to higher dimensional systems. In this regard, some mention will be made of generalizations to higher dimensional systems which tend to require computer simulation methods [7].

Both acoustic [7, 8] and optical metamaterials [11, 43] are treated. Though these involve different physical theories, many of the ideas related in the design of metamaterials are common to optical and acoustic metamaterials. First, a treatment of acoustic metamaterials is given, followed by that of optical metamaterials. After these, a general discussion of the applications of metamaterials is presented.

5.2.1   NEGATIVE EFFECTIVE MASS

In this section a study is presented of a one-dimensional model of an acoustic metamaterial exhibiting a negative effective mass [8, 10, 12]. The model studied [12] is simple to treat theoretically and facilitates the presentation of the basic ideas behind resonant effects leading to negative effective mass metamaterials. In technological applications, however, designed systems exhibiting greater sophistication would generally be developed to meet the detailed requirements of a device technology. Such generalizations, nevertheless, would be elaborations of the basic structures and ideas presented here. In addition, the one-dimensional metamaterial model in the present study can easily be extended to high-dimensional metamaterials exhibiting negative effective mass [13].

Consider the one-dimensional metamaterial design [12] presented schematically in Fig. 5.2. It is a generalization of the basic infinite one-dimensional mass chain studied in solid state physics [1, 2]. In the present model shells of mass m_1 are constrained to move along the horizontal x-axis and are coupled to one another by nearest neighbor massless springs of spring constant k_1. Within each of the shells is a point mass m_2 which is constrained to move along the horizontal x-axis and is coupled to its enclosing shell by a massless spring of spring constant k_2. The basic resonator of the system is the mass shell and its enclosed mass, and the resonance arises from the relative harmonic motions of the mass shell and the mass which it contains [12].

In this regard, the model is of an infinite chain of coupled resonators [12]. Along the chain the resonators are labeled by integers n so that in equilibrium the center of the nth shell is located on the x-axis at $x_n = nL$ where L is the nearest neighbor separation between sites in the undistorted chain. Similarly, in the equilibrium configuration of the chain the enclosed mass within the nth shell is also located at $x_n = nL$. Continuing to the treatment of the system dynamics, the displacement from equilibrium of the center of the nth shell is $u_1^{(n)}$ and the displacement from equilibrium of the enclosed mass in the nth shell is $u_2^{(n)}$.

With this notation it then follows that the dynamical equations for the model in Fig. 5.2 are [12]:

$$m_1 \frac{d^2 u_1^{(n)}}{dt^2} + k_1 \left(2u_1^{(n)} - u_1^{(n+1)} - u_1^{(n-1)} \right) + k_2 \left(u_1^{(n)} - u_2^{(n)} \right) = 0 \tag{5.5a}$$

$$m_2 \frac{d^2 u_2^{(n)}}{dt^2} + k_2 \left(u_2^{(n)} - u_1^{(n)} \right) = 0. \tag{5.5b}$$

Here m_1 is the mass of a shell, m_2 is the mass of the mass enclosed by a shell, k_1 is the spring constant between nearest neighbor shells, and k_2 is the spring constant between an enclosed mass and its enclosing shell. In this regard, note that in the limit that $k_2 \to 0$ Eq. (5.5a) reduces to the equation of motion of a chain composed of masses m_1. In addition, from Eq. (5.5b) it is seen that the natural frequency of the vibrational motion of an enclosed mass relative to its fixed enclosing shell is $\omega_2 = \sqrt{\frac{k_2}{m_2}}$.

The general form of the vibrational solutions of the modes of Eqs. (5.5) are expressed as

$$u_\gamma^{(n)} = \tilde{u}_\gamma e^{i(qx_n - \omega t)}, \tag{5.6}$$

where $\gamma = 1, 2$. Upon substituting Eq. (5.6) into Eqs. (5.5) the equations of motion are resolved into an algebraic eigenvalue problem for the frequency, ω, and its corresponding eigenvectors. This is given by [12]

$$\begin{vmatrix} -m_1\omega^2 + 2k_1 (1 - \cos qL) + k_2 & -k_2 \\ -k_2 & -m_2\omega^2 + k_2 \end{vmatrix} \begin{bmatrix} u_1^{(n)} \\ u_2^{(n)} \end{bmatrix} = 0. \tag{5.7}$$

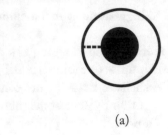

(a)

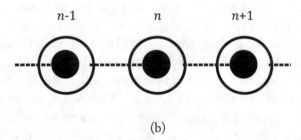

(b)

Figure 5.2: One-dimensional negative effective mass system. In (a) is the basic resonator unit composed of a mass shell of mass m_1 which encloses a mass m_2. The shell and enclosed mass are connected by a spring of spring constant k_2. In (b) an infinite chain of resonators are nearest neighbor coupled by springs of spring constant k_1. The sites of the chain are labeled by integers $\ldots, n-1, n, n+1, \ldots$. The chain is shown configured in its equilibrium configuration. The x-axis is the horizontal and along the axis of the chain.

The modal dispersion relation of the chain of resonators is obtained from the determinant of the matrix in Eq. (5.7) and the application of periodic boundary conditions to set the values of q supported by the chain. From the calculation of the determinate in Eq. (5.7), the conditions for modal solutions to exist is that

$$\eta^4 - \left[1 + \frac{m_2}{m_1} + 2\frac{k_1}{k_2}\frac{m_2}{m_1}(1 - \cos qL)\right]\eta^2 + 2\frac{k_1}{k_2}\frac{m_2}{m_1}(1 - \cos qL) = 0, \qquad (5.8a)$$

where $\eta = \frac{\omega}{\omega_2} = \omega\sqrt{\frac{m_2}{k_2}}$. From the application of the periodic boundary conditions between the ends of the infinite chain to the solutions in Eq. (5.6), the values of q in Eqs. (5.7) and (5.8a) are restricted to the set

$$q = \frac{n\pi}{NL}, \qquad (5.8b)$$

where $n = 0, \pm 1, \pm 2, \ldots, \pm N$. Combining Eqs. (5.8a) and (5.8b) then determines the modal solutions of the chain.

The determinant of the matrix in Eq. (5.7) can be written in an alternative form as [12]

$$\omega^2 = 2 \frac{k_1}{m_{eff}(\omega)} (1 - \cos qL) \tag{5.9a}$$

where

$$m_{eff}(\omega) = m_1 + m_2 + \frac{m_2 \dfrac{\omega^2}{\omega_2^2}}{1 - \dfrac{\omega^2}{\omega_2^2}} \tag{5.9b}$$

is a frequency-dependent effective mass. Equations (5.9) have the same solutions as Eq. (5.8a) but are written in the form of the dispersion relation of a chain of mass m_{eff} and spring constant k_1. The effective mass, however, in this alternative form of the chain dispersion relation is now a function of the frequency. In this sense, the dispersion relation of the chain of coupled resonators can be regarded as a simple mass chain with frequency-dependent masses.

An important point to note from Eq. (5.9b) is that $m_{eff}(\omega)$ becomes negative in the range $\omega_2 < \omega < \sqrt{\frac{m_1+m_2}{m_1}} \omega_2 = \sqrt{\frac{m_1+m_2}{m_1 m_2} k_2}$ where the upper limit of the interval is the vibrational frequency of the $m_1 - m_2$ shell enclosed mass unit [12]. For this set of frequencies, the model exhibits the physics of a vibrational chain composed of negative masses. Notice, however, for a negative effective mass in Eq. (5.9a), q must be imaginary. In this regard, the model displays new metamaterial features arising from its engineered structure. For example, in a system exhibiting negative mass, the acceleration is opposite the direction of the force. This is not commonly found in nature and is a direct consequence of the properties of the shell resonators of the chain. Any modal solutions falling in the regions of negative mass will display the unusual characteristics associated with the new metamaterial physics [12].

Negative Effective Mass: Two Dimensional Model

The model for a one-dimensional chain with a negative effective mass can be generalized in a straightforward manner to treat higher dimensional lattices [7, 8, 13]. For example, in two-dimensions this is done by placing sets of mass shell resonators coupled by nearest neighbor spring couplings to the sites of a two-dimensional lattice. A similar treatment on a three-dimensional lattice applies to models of three-dimensional metamaterials. In the following, an outline of the theory of a highly simplistic two-dimensional negative mass acoustic metamaterials is presented.

Consider a square lattice with lattice sites [13] labeled by integers (n, m). The space coordinates of the lattice points are expressed in terms of the integers of the site labels by $(x_n, y_m) = (Ln, Lm)$ where L is the nearest neighbor site separation on the lattice. As in the one-dimensional metamaterial model, shell resonators are placed on each lattice site, and each resonator is composed as a shell of mass m_1 which is spring coupled to an enclosed mass of mass

m_2. Now, however, there are two springs coupling the enclosed mass to the mass shell. There is a spring coupling k_{2x} between the enclosed mass to the shell for a harmonic motion in the x-direction and a spring coupling k_{2y} between the enclosed mass to the shell for a harmonic motion in the y-direction. The resonators are in turn coupled to one another by nearest neighbor springs along the x- and y-directions. The nearest neighbor couplings between resonators in the x-direction are denoted k_{1x} and the nearest neighbor couplings in the y-directions are denoted k_{1y}. (For a more detailed treatment of this model, including shear forces, the reader is referred to [13]. Here only the simplest treatment is indicated.)

If the displacement of the shell at the site (n, m) is denoted by $\left(u_{1x}^{(n,m)}, u_{1y}^{(n,m)}\right)$ and the displacement of the enclosed mass at the site (n, m) is denoted by $\left(u_{2x}^{(n,m)}, u_{2y}^{(n,m)}\right)$, then the equations of motion along the two spatial axis are given by [13]

$$m_1 \frac{d^2 u_{1x}^{(n,m)}}{dt^2} + k_{1x}\left(2u_{1x}^{(n,m)} - u_{1x}^{(n+1,m)} - u_{1x}^{(n-1,m)}\right) + k_{2x}\left(u_{1x}^{(n,m)} - u_{2x}^{(n,m)}\right) = 0, \quad (5.10a)$$

$$m_2 \frac{d^2 u_{2x}^{(n,m)}}{dt^2} + k_{2x}\left(u_{2x}^{(n,m)} - u_{1x}^{(n,m)}\right) = 0, \quad (5.10b)$$

$$m_1 \frac{d^2 u_{1y}^{(n,m)}}{dt^2} + k_{1y}\left(2u_{1y}^{(n,m)} - u_{1y}^{(n,m+1)} - u_{1y}^{(n,m-1)}\right) + k_{2y}\left(u_{1y}^{(n,m)} - u_{2y}^{(n,m)}\right) = 0, \quad (5.10c)$$

$$m_2 \frac{d^2 u_{2y}^{(n,m)}}{dt^2} + k_{2y}\left(u_{2y}^{(n,m)} - u_{1y}^{(n,m)}\right) = 0. \quad (5.10d)$$

The modal solutions of these equations are of a generalized form of Eq. (5.6) given by [13]

$$u_{\gamma j}^{(n,m)} = \tilde{u}_{\gamma j} e^{i\left(q_x x_n + q_y y_m - \omega t\right)}, \quad (5.11)$$

where $\gamma = 1, 2$ and $j = x, y$. Substituting Eq. (5.11) into Eqs. (5.10) yields an algebraic eigenvalue problem for the modal frequencies and modal eigenfunctions. Developing the solutions of the algebraic problem in a manner which parallels that of the one-dimensional systems yields a simple two-dimensional vibrational system of effective masses. In general, the effective masses are found to be different for both x- and y-motions.

The solutions for the dispersion relation of the system of resonant shells [13] can be rewritten into the form of an effective medium model similar to that treated in Eqs. (5.8) and (5.9). From these considerations two effective masses are extracted, one for each of the two axes of the system. The effective dispersion relation is obtained from the form [13]

$$\left[\omega^2 - 2\frac{k_{1x}}{m_{eff,x}(\omega)}\left(1 - \cos q_x L\right)\right]\left[\omega^2 - 2\frac{k_{1y}}{m_{eff,y}(\omega)}\left(1 - \cos q_y L\right)\right] = 0, \quad (5.12)$$

where

$$m_{eff,x}\left(\omega\right) = m_1 + m_2 + \frac{m_2 \dfrac{\omega^2}{\omega_{2x}^2}}{1 - \dfrac{\omega^2}{\omega_{2x}^2}} \tag{5.13a}$$

along the x-axis, and

$$m_{eff,y}\left(\omega\right) = m_1 + m_2 + \frac{m_2 \dfrac{\omega^2}{\omega_{2y}^2}}{1 - \dfrac{\omega^2}{\omega_{2y}^2}} \tag{5.13b}$$

along the y-axis where $\omega_{2x} = \sqrt{\frac{k_{2x}}{m_2}}$ and $\omega_{2y} = \sqrt{\frac{k_{2y}}{m_2}}$. Both of the effective masses in Eqs. (5.12) and (5.13) are found to exhibit regions of negative effective mass [13].

5.2.2   NEGATIVE YOUNG'S MODULUS

Another acoustic metamaterial design is associated with the interesting property of negative Young's modulus [41]. This too can be made to exist in a properly engineered metamaterial [7, 8, 40, 41]. For a simple treatment a one-dimensional model is studied which facilitates the presentation of the basic ideas behind resonant effects leading to negative Young's moduli metamaterials. As with the earlier discussions of negative mass, only the simplest model of a negative Young's moduli metamaterial is presented in the following treatment. In technological applications practical systems of greater sophistication are often employed to meet the requirements of a device technology.

Consider the one-dimensional metamaterial [41] design presented schematically in Fig. 5.3. In the present model masses of mass m_1 are constrained to move along the horizontal x-axis and are coupled to one another by nearest neighbor massless springs of spring constant k_1. This arrangement of masses and springs forms a basic one-dimensional mass chain which is commonly studied in solid state physics. To this chain, however, are added a series of massless, rigid, trusses which are free to rotate at the sites of the chain masses to which they are attached and at their off-chain truss vertices. At the off-chain vertices are connected masses m_3 which are constrained to move along the vertical y-axis relative to the truss vertices. The masses m_3 are connected to the truss off-chain vertices by massless springs of spring constant k_3. The basic resonator of the system is now the truss off-chain mass system [41].

In this regard, the model is of an infinite chain coupled to off-chain trusses, and these trusses are in turn coupled to off-chain mass attachments [41]. Along the chain the m_1 masses are labeled by integers n so that in equilibrium the nth mass m_1 is located on the x-axis at $x_n = nL$ where L is the nearest neighbor separation between sites in the undistorted chain.

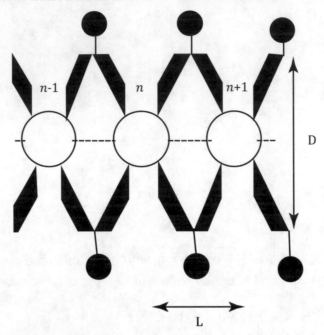

Figure 5.3: Negative Young's Modulus chain. Here the masses, m_2, and springs, k_2, within the chain in Fig. 5.2 have been removed and trusses have been added above and below the chain. Masses m_3 are attached to the truss upper-most and lower-most vertices by springs of spring constants k_3. The chain is shown configured in its equilibrium configuration. The x-axis is horizontal and along the axis of the chain. The y-axis is vertical and perpendicular to the axis of the chain.

In the equilibrium configuration the nth off-chain truss vertices are located at $\left(x_n + \frac{L}{2}, \pm y_n\right)$ where $y_n = \frac{D}{2}$, and the nth off-chain masses m_3 are located at $\left(x_n + \frac{L}{2}, \pm y_n \pm d\right)$.

When the system is displaced from equilibrium, the displacement of the nth mass m_1 from equilibrium is $u_1^{(n)}$. The displacement from equilibrium along the y-axis of the upper nth mass m_3 is $v_3^{(n)}$, and the displacement from equilibrium along the y-axis of the lower nth mass m_3 is $-v_3^{(n)}$. Similarly, the off-chain truss vertices are displaced from equilibrium. The displacement from equilibrium along the y-axis of the upper nth truss vertex is $v_1^{(n)}$, and the displacement from equilibrium along the y-axis of the lower nth truss vertex is $-v_1^{(n)}$. It should be noted that only modes which are symmetric in displacements along the y-axis are considered in this formulation.

With this notation it then follows that the dynamical equations for the model in Fig. 5.3 are [41]:

$$m_1 \frac{d^2 u_1^{(n)}}{dt^2} + k_1 \left(2u_1^{(n)} - u_1^{(n+1)} - u_1^{(n-1)}\right)$$

$$+k_3 \left(v_1^{(n)} - v_3^{(n)}\right)\frac{L}{D} - k_3 \left(v_1^{(n-1)} - v_3^{(n-1)}\right)\frac{L}{D} = 0 \qquad (5.14a)$$

$$m_3 \frac{d^2 v_3^{(n)}}{dt^2} + k_3 \left(v_3^{(n)} - v_1^{(n)}\right) = 0. \qquad (5.14b)$$

Here m_1 is a chain mass coupled along the chain, m_3 is the mass of the off-chain mass attached to the off-chain truss vertices, k_1 is the spring constant of the couplings along the chain, and k_3 is the spring constant between the off-chain masses and their off-chain truss vertices. In this regard, note that in the limit that $k_3 \to 0$ Eq. (5.14a) reduces to the equation of motion of a chain composed of masses m_1. In addition, from Eq. (5.14b) it is seen that the natural frequency of the vibrational motion of a mass m_3 attacked to the truss is $\omega_3 = \sqrt{\frac{k_3}{m_3}}$. The factors of $\frac{L}{D}$ arise from the transmission of the forces from the m_3 masses onto the m_1 masses through the massless trusses.

Due to the rigid nature of the trusses, the displacements $u_1^{(j)}$ can be related to the displacements $v_1^{(j)}$. Consider the triangular truss with the y-displacement $v_1^{(j)}$ at its off-chain vertex. The x-displacements of the truss at the chain are $u_1^{(j)}$ and $u_1^{(j+1)}$ so that the base of the equilateral triangle of the truss changes from a length L to a new length $L + u_1^{(j+1)} - u_1^{(j)}$. To the leading ordering in the small displacements $u_1^{(j)}$ and $v_1^{(j)}$, the requirement that the truss legs are rigid yields the relationship [41]

$$v_1^{(j)} = -\frac{L}{2D}\left(u_1^{(j+1)} - u_1^{(j)}\right). \qquad (5.15)$$

Applying Eq. (5.15) in Eqs. (5.14) to eliminate the $v_1^{(j)}$ and taking [41]

$$u_1^{(n)} = \tilde{u}_1 e^{i(qx_n - \omega t)} \qquad (5.16a)$$

$$v_3^{(n)} = \tilde{v}_3 e^{i(qx_n - \omega t)} \qquad (5.16b)$$

as the general form of the solution, reduces Eqs. (5.14) to an algebraic eigenvalue problem for ω and $u_1^{(n)}$, $v_3^{(n)}$. This is expressed in terms of the homogeneous matrix equation [41]

$$\begin{vmatrix} -m_1\omega^2 + 2k_1(1 - \cos qL) & m_3\omega^2\frac{L}{D}\left(e^{-iqL} - 1\right) \\ \frac{1}{2}k_3\frac{L}{D}\left(e^{iqL} - 1\right) & -m_3\omega^2 + k_3 \end{vmatrix} \begin{vmatrix} u_1^{(n)} \\ v_3^{(n)} \end{vmatrix} = 0 \qquad (5.17)$$

and its solvability conditions.

The modal dispersion relation of the chain is obtained from the determinant of the matrix in Eq. (5.17) along with the application of periodic boundary conditions for the determination of the q supported by the chain [41]. In this way it follows from Eq. (5.17) that the frequency is given in terms of q by the solutions of

$$\eta^4 - \left[1 + 2\frac{k_1}{m_1}\frac{m_3}{k_3}\left(1 + \frac{1}{2}\frac{k_3}{k_1}\mu\right)(1 - \cos qL)\right]\eta^2 + 2\frac{k_1}{k_3}\frac{m_3}{m_1}(1 - \cos qL) = 0, \qquad (5.18a)$$

where $\eta = \frac{\omega}{\omega_3} = \omega\sqrt{\frac{m_3}{k_3}}$ and $\mu = \frac{L^2}{D^2}$. From the application of the periodic boundary conditions between the ends of the infinite chain to the solutions in Eqs. (5.16), the values of q in Eqs. (5.17) and (5.18a) are restricted to the set

$$q = \frac{n\pi}{NL}, \qquad (5.18b)$$

where $n = 0, \pm 1, \pm 2, \ldots, \pm N$. Combining Eqs. (5.18a) and (5.18b) then completely determines the modal solution of the chain.

The determinant of the matrix in Eq. (5.17) can be written in an alternative form as [17, 41]

$$\omega^2 = 2\frac{k_{eff}(\omega)}{m_1}(1 - \cos qL) \qquad (5.19a)$$

where

$$k_{eff}(\omega) = k_1 - \frac{1}{2}\frac{k_3\mu\dfrac{\omega^2}{\omega_3^2}}{1 - \dfrac{\omega^2}{\omega_3^2}} \qquad (5.19b)$$

is a frequency-dependent effective spring constant. Equations (5.19) have the same solutions as Eq. (5.18a) but are written in the form of the dispersion relation of a chain of mass m_1 and spring constant k_{eff}. The effective spring constant, however, in this alternative form of the chain dispersion relation is now a function of the frequency. In this sense, the dispersion relation of the chain and trusses can be regarded as a simple mass chain with frequency-dependent spring constants [41].

An important point to note from Eq. (5.19b) is that $k_{eff}(\omega)$ becomes negative in the interval [17, 41] $\sqrt{\dfrac{1}{1 + \frac{1}{2}\mu\frac{k_3}{k_1}}}\omega_3 = \sqrt{\dfrac{k_3}{m_3\left(1 + \frac{1}{2}\mu\frac{k_3}{k_1}\right)}} < \omega < \omega_3$ where the lower limit of the range is the vibrational frequency of the truss-mass unit. For this set of frequencies, the physics of the model is that of a vibrational chain composed of a negative effective spring constant and the corresponding values of q in Eq. (5.19a) must be imaginary. One consequence of this spring constant design feature is that the Young's modulus is negative [41]. In this regard, it shall be

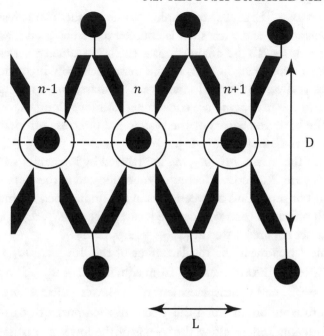

Figure 5.4: Negative Young's modulus and negative effective mass chain. The model is that in Fig. 5.3 with the springs and masses within the mass shells reintroduced into the shells of the chain. The chain is shown configured in its equilibrium configuration. The x-axis is horizontal and along the axis of the chain. The y-axis is vertical and perpendicular to the axis of the chain.

seen later that a response of this type has important device applications when combined with the negative mass design mechanisms treated earlier. The combination of these two engineered responses is now addressed [41].

5.2.3   NEGATIVE YOUNG'S MODULUS AND NEGATIVE EFFECTIVE MASS

Another metamaterial design [7, 8] is associated with the interesting properties of a system exhibiting both a negative Young's modulus and a negative effective mass [14, 15, 17, 41]. This can be made to exist in an engineered metamaterial containing both mass shell- and truss-based resonators. These resonator features will now be introduced into the treatment of a system based on the one-dimensional mass chain models. As in the earlier discussions of resonant acoustic metamaterials, the ideas presented here, in one dimensions, are generalizable to higher dimensional systems and can also be modified to meet the requirements of a device technology.

Consider the one-dimensional metamaterial [14, 15, 17, 41] design presented in Fig. 5.4. In the model, mass shells of mass m_1 and their enclosed masses m_2 are constrained to move along the horizontal x-axis. The shells are coupled to one another by nearest neighbor massless springs of spring constant k_1 and each shell is coupled to its enclosed mass m_2 by a massless spring of spring constant k_2. To the chain of mass shells are added a series of massless, rigid, trusses which are free to rotate at the sites of the mass shells to which they are attached and at the off-chain truss vertices. At the off-chain vertices masses m_3 are attached by springs of spring constant k_3, and the masses m_3 are constrained to move along the y-axis relative to the off-chain vertices. The basic resonators of the system are the truss off-chain mass system and the mass shell enclosed mass system [17, 41].

Along the chain the masses m_1, m_2, m_3 are labeled by integers n so that in equilibrium the nth masses m_1, m_2 are, respectively, centered on or located at the x-axis at $x_n = nL$ where L is the nearest neighbor separation between sites in the undistorted chain. In the equilibrium configuration the nth off-chain truss vertices are located at $\left(x_n + \frac{L}{2}, \pm y_n\right)$ where $y_n = \frac{D}{2}$, and the nth off-chain masses m_3 are located at $\left(x_n + \frac{L}{2}, \pm y_n \pm d\right)$.

In the dynamical treatment the displacement of the nth mass m_1 from equilibrium is $u_1^{(n)}$ and the displacement of the nth mass m_2 from equilibrium is $u_2^{(n)}$. The displacement of the upper nth mass m_3 is $v_3^{(n)}$, and the displacement of the lower nth mass m_3 is $-v_3^{(n)}$. Similarly, the displacement from equilibrium along the y-axis of the upper nth truss vertex is $v_1^{(n)}$, and the displacement from equilibrium along the y-axis of the lower nth truss vertex is $-v_1^{(n)}$. As in the earlier considerations only modes which are symmetric in displacements along the y-axis are considered in this formulation.

It then follows that the dynamical equations for the model in Fig. 5.4 are [14, 15, 17, 41]:

$$m_1 \frac{d^2 u_1^{(n)}}{dt^2} + k_1 \left(2u_1^{(n)} - u_1^{(n+1)} - u_1^{(n-1)}\right) + k_2 \left(u_1^{(n)} - u_2^{(n)}\right)$$

$$+ k_3 \left(v_1^{(n)} - v_3^{(n)}\right) \frac{L}{D} - k_3 \left(v_1^{(n-1)} - v_3^{(n-1)}\right) \frac{L}{D} = 0 \qquad (5.20a)$$

$$m_2 \frac{d^2 u_2^{(n)}}{dt^2} + k_2 \left(u_2^{(n)} - u_1^{(n)}\right) = 0, \qquad (5.20b)$$

$$m_3 \frac{d^2 v_3^{(n)}}{dt^2} + k_3 \left(v_3^{(n)} - v_1^{(n)}\right) = 0. \qquad (5.20c)$$

These equations [17] are essentially a linear merger of the different resonant force terms in Eqs. (5.5) and (5.14), i.e., the terms involving m_2, k_2, m_3, k_3 are carried over and combined. In this regard, note that in the limit that k_2, $k_3 \to 0$ Eq. (5.20a) reduces to the equation of motion of a chain composed of masses m_1. In addition, from Eqs. (5.20b) and (5.20c) it is seen that the natural frequency of the vibrational motion of an enclosed mass relative to its fixed enclosing shell is $\omega_2 = \sqrt{\frac{k_2}{m_2}}$, and the natural frequency of the vibrational motion of one of the masses attached to the truss is $\omega_3 = \sqrt{\frac{k_3}{m_3}}$. The factors of $\frac{L}{D}$ arise from the transmission of the forces from the m_3 masses onto the m_1 masses by the massless trusses.

Following the earlier discussion for the system of trusses leading to negative effective spring constants, the truss motions $v_1^{(j)}$ can be related to the chain displacements $u_1^{(j)}$. To leading ordering in the small displacements $u_1^{(j)}$ and $v_1^{(j)}$, the relationship [17]

$$v_1^{(j)} = -\frac{L}{2D}\left(u_1^{(j+1)} - u_1^{(j)}\right) \tag{5.21}$$

is again obtained from the requirement that the truss legs are rigid.

Applying Eq. (5.21) in Eqs. (5.20) to illuminate the $v_1^{(j)}$ and taking [14, 15, 17, 41]

$$u_\gamma^{(n)} = \tilde{u}_\gamma e^{i(qx_n - \omega t)} \quad \text{for} \quad \gamma = 1,2 \tag{5.22a}$$

$$v_3^{(n)} = \tilde{v}_3 e^{i(qx_n - \omega t)} \tag{5.22b}$$

as the forms of the general modal solutions reduces Eqs. (5.20) to an algebraic eigenvalue problem for ω and $u_1^{(n)}, u_2^{(n)}, v_3^{(n)}$. This is given by the homogeneous matrix equation [14, 15, 17]

$$\begin{vmatrix} -m_1\omega^2 + 2k_1\left(1 - \cos qL\right) & -m_2\omega^2 & m_3\omega^2\left(e^{-iqL} - 1\right)\frac{L}{D} \\ -k_2 & -m_2\omega^2 + k_2 & 0 \\ \frac{1}{2}k_3\left(e^{iqL} - 1\right)\frac{L}{D} & 0 & -m_3\omega^2 + k_3 \end{vmatrix} \begin{vmatrix} u_1^{(n)} \\ u_2^{(n)} \\ v_3^{(n)} \end{vmatrix} = 0. \tag{5.23}$$

The modal dispersion relation of the chain is again obtained from the determinant of the matrix in Eq. (5.23) along with the application of periodic boundary conditions for the determination of the q supported by the chain. In this way it follows from Eq. (5.23) that the frequency is given in terms of q by the solutions of [17]

$$\eta^6 - \left[1 + \frac{k_3 m_2}{m_3 k_2} + \frac{k_1}{m_1}\frac{m_2}{k_2}\left[\frac{k_2}{k_1} + 2\left(1 + \frac{1}{2}\frac{k_3}{k_1}\mu\right)(1 - \cos qL)\right]\right]\eta^4$$

$$+ \left[\frac{m_2 k_3}{k_2 m_3}\left(1 + \frac{m_2}{m_1}\right) + 2\frac{k_1 m_2}{m_1 k_2}\left(1 + \frac{k_3 m_2}{m_3 k_2} + \frac{1}{2}\frac{k_3}{k_1}\mu\right)(1 - \cos qL)\right]\eta^2$$

$$-2\frac{k_1}{m_1}\frac{k_3}{m_3}\frac{m_2^2}{k_2^2}(1 - \cos qL) = 0 \tag{5.24a}$$

where $\eta = \frac{\omega}{\omega_2} = \omega\sqrt{\frac{m_2}{k_2}}$ and $\mu = \frac{L^2}{D^2}$.

From the application of the periodic boundary conditions between the ends of the infinite chain to the solutions in Eqs. 5.22, the values of q in Eqs. (5.23) and (5.24a) are restricted to the set

$$q = \frac{n\pi}{NL} \tag{5.24b}$$

where $n = 0, \pm 1, \pm 2, \ldots, \pm N$. Combining Eqs. (5.24a) and (5.24b) then determines the modal solution of the chain.

The determinant of the matrix in Eq. (5.23) can be alternatively expressed in the form [17]

$$\omega^2 = 2 \frac{k_{eff}(\omega)}{m_{eff}(\omega)} (1 - \cos qL) \tag{5.25a}$$

where

$$m_{eff}(\omega) = \frac{m_1 m_2 \omega^2 - (m_1 + m_2) k_2}{m_2 \omega^2 - k_2} = m_1 + m_2 + \frac{m_2 \omega^2}{\omega_2^2 - \omega^2} \tag{5.25b}$$

$$k_{eff}(\omega) = \frac{1}{2} \frac{k_3 m_3 \mu \omega^2 + 2 k_1 m_3 \omega^2 - 2 k_1 k_3}{m_3 \omega^2 - k_3} = k_1 - \frac{1}{2} \frac{k_3 \mu \omega^2}{\omega_3^2 - \omega^2} \tag{5.25c}$$

are the frequency-dependent effective mass and spring constant. These effective mass and spring [14, 15, 17] constant parameters are the same as those obtained in the two earlier models studied for negative effective mass and Young's constants. The region of negative effective mass is given by [17]

$$\omega_2 = \sqrt{\frac{k_2}{m_2}} < \omega < \sqrt{\frac{m_1 + m_2}{m_1}} \omega_2 = \sqrt{\frac{m_1 + m_2}{m_1} \frac{k_2}{m_2}} \tag{5.26a}$$

and the region of negative effective spring constant or Young's Modulus is [17]

$$\sqrt{\frac{1}{1 + \frac{1}{2}\mu\frac{k_3}{k_1}}} \omega_3 = \sqrt{\frac{1}{1 + \frac{1}{2}\mu\frac{k_3}{k_1}} \frac{k_3}{m_3}} < \omega < \omega_3 = \sqrt{\frac{k_3}{m_3}}. \tag{5.26b}$$

It is seen from Eqs. (5.26) that the parameters of the system can be chosen so that there are regions of frequency in which the effective mass and effective spring constant are both negative. It shall be found later that this region of negative effective mass and spring constants displays a variety of interesting new physical properties for its corresponding propagating modes. These interesting new properties are of particular importance in the long wavelength limit in which the system behaves as a homogeneous metamaterial.

Double-Negative Properties

It is interesting to look at the properties of the effective medium representation of the chain in Fig. 5.4 with a focus on the long wavelength limit [17]. This is the limit in which the system behaves as a uniform homogeneous medium in a continuum representation. The region of interest in this regard is the set of frequencies at which the chain gives the so-called double negative response. In this response the material displays simultaneously a negative effective mass and a negative effective Young's modulus. The properties exhibited by the metamaterial in a double-negative response are atypical of media commonly found in nature. The double-negative response arises from the resonant properties of the mass-shell-enclosed-mass and the truss resonators that have been engineered into the body of the metamaterial [14, 15].

To understand the nature of this response, consider Eqs. (5.22) and (5.25) in the $q \to 0$ limit. In this limit it follows that the dispersion relation of the effective medium becomes [14, 15, 17, 41]

$$\omega^2 = \frac{k_{eff}(\omega)}{m_{eff}(\omega)} q^2 L^2,$$ (5.27a)

and the corresponding continuum limit representation of the wavefunction is of the form

$$u_1(x,t) = \tilde{u}_1 e^{i(qx - \omega t)}.$$ (5.27b)

From Eq. (5.27a) the continuum limit wave equation for the chain can be written as a differential equation involving the wavefunction in Eq. (5.27b). Written in this format the effective medium wave equation is [17]

$$E_{eff} \frac{d^2}{dx^2} u(x,t) = -\rho_{eff}\omega^2 u(x,t).$$ (5.28)

Here the mass density of the chain is defined by

$$\rho_{eff} = \frac{m_{eff}}{LA}$$ (5.29a)

and the Young's modulus of the chain by

$$E_{eff} = \frac{k_{eff} L}{A}$$ (5.29b)

where A is an effective chain area in the perpendicular plane to the axis of the chain. In regard to the definition of the area A, a crystal can be built by stacking parallel chains so that a natural choice for A would be the cross-sectional area per chain in an infinite three-dimensional crystal array of chains.

From Eq. (5.28) and the definitions in Eqs. (5.29) it follows that the dispersion relation of the planewave modes can be written as $E_{eff} q^2 = \rho_{eff}\omega^2$. The phase velocity of the mode in Eq. (5.27b) is then obtained from this form of the dispersion relation as [14, 15, 17, 41]

$$v_p = \frac{\omega}{q} = \frac{q}{|q|} \sqrt{\frac{E_{eff}}{\rho_{eff}}},$$ (5.30a)

and it is seen from Eqs. (5.30) that the phase velocity is parallel to the direction of q.

A point to note [17] is that the phase velocity is real, with propagating wave solutions, for both of the cases $\rho_{eff}, E_{eff} \geq 0$ or $\rho_{eff}, E_{eff} \leq 0$. For the remaining combinations of ρ_{eff} and E_{eff}, however, the solutions have complex phase velocities which yield non-propagating excitations in the system. These solutions exponentially decay or increase in space. Only the propagating solutions are of interest in applications considered later.

The propagation of energy in the system is governed by the group velocity. This, in general, is different from the phase velocity in Eq. (5.30a), particularly in systems with nonlinear dispersion relations. In most elementary mechanics texts, it is shown that the group velocity is given by [14, 15, 17]

$$v_g = \frac{\partial \omega}{\partial q} \tag{5.30b}$$

and that it is related to the propagation along the chain of the envelope of a localized pulse. Consequently, given the dispersion relation of the modes in the system, the group velocity can be readily computed.

The partial $\frac{\partial \omega}{\partial q}$ along the effective medium chain is obtained directly from the dispersion relation in Eq. (5.27a) rewritten in the form [17]

$$D(\omega, q) = m_{eff}\omega^2 - k_{eff}(\omega) q^2 L^2 = 0. \tag{5.31}$$

Using the standard calculus relationship along the surfaces of constant $D(\omega, q)$, it is found that

$$dD = \frac{\partial D}{\partial \omega} d\omega + \frac{\partial D}{\partial q} dq = 0 \tag{5.32a}$$

so that

$$\frac{\partial \omega}{\partial q} = -\frac{\dfrac{\partial D}{\partial q}}{\dfrac{\partial D}{\partial \omega}}. \tag{5.32b}$$

From Eq. (5.31) it then follows that the partial $\frac{1}{LA}\frac{\partial D}{\partial q} = -2E_{eff}q$ and again from Eq. (5.31) that the partial $\frac{1}{LA}\frac{\partial D}{\partial \omega} = 2\rho_{eff}\omega + \left(\frac{\partial \rho_{eff}}{\partial \omega} - \frac{\rho_{eff}}{E_{eff}}\frac{\partial E_{eff}}{\partial \omega}\right)\omega^2$. Consequently, from Eqs. (5.30)–(5.32) the group velocity is [15, 17]

$$v_g = \frac{2E_{eff}}{2\rho_{eff}\omega + \left(\dfrac{\partial \rho_{eff}}{\partial \omega} - \dfrac{\rho_{eff}}{E_{eff}}\dfrac{\partial E_{eff}}{\partial \omega}\right)\omega^2} q. \tag{5.33}$$

From Eqs. (5.25b) and (5.25c) and the definitions in Eqs. (5.29a) and (5.29b) the partials in Eq. (5.33) are evaluated. This gives [17]

$$\frac{\partial \rho_{eff}}{\partial \omega} = 2\frac{m_2}{LA}\frac{\omega\omega_2^2}{(\omega_2^2 - \omega^2)^2} \tag{5.34}$$

and

$$\frac{\partial E_{eff}}{\partial \omega} = -\frac{L}{A} k_3 \mu \frac{\omega \omega_3^2}{\left(\omega_3^2 - \omega^2\right)^2}.$$ (5.35)

It follows then from Eq. (5.33)–(5.35) that in the case that $\rho_{eff}, E_{eff} \geq 0$ (recall that $m_2, k_3 \geq 0$) both the numerator and denominator in Eq. (5.33) are positive so that the phase velocity and group velocities are in the same direction. This arises as the coefficient in front of q in Eq. (5.33) is positive [17].

In the case [17], however, that $\rho_{eff}, E_{eff} \leq 0$ (recall that $m_2, k_3 \geq 0$) the denominator in Eq. (5.33) is positive, but the numerator is negative. In this case the phase velocity and group velocities are in opposite directions. This arises as the coefficient in front of q in Eq. (5.33) is now negative.

It is seen that for $\rho_{eff}, E_{eff} \leq 0$ planewaves propagate along the chain but the flow of energy in now opposite the direction of the phase velocity of the planewaves. This is an anomalous property that exists in the system due to the engineered resonant features in the system [17]. It requires both ρ_{eff} and E_{eff} to be negative which generally is not found in naturally occurring media in the long wavelength homogeneous limit. Later it shall be shown that a number of interesting physical properties arise in metamaterials for the case in which $\rho_{eff}, E_{eff} \leq 0$.

Next, it shall be discussed how similar new properties to those of double negative acoustic metamaterials can be introduced into the dynamics of electromagnetic systems. This is again done through the ideas of metamaterials formulated as particular types of composites involving the introduction of engineered resonant features. In this regard, it is shown that in properly engineered dielectric metamaterials it is possible to make a medium in which the flow of energy (i.e., the group velocity) is opposite the phase velocity of propagating waves. The ability to design such optical materials has opened new fields of optical technology with novel device applications. These optical metamaterials are now discussed.

Following the discussions of optical metamaterials, some of the interesting consequence of the physics of media in which the phase velocity of the excitations is opposite the group velocity are treated. These include novel refraction effects in the passage of excitations between different acoustic and optical media, super lensing effects, sensor applications, and cloaking mechanisms. Most of these ideas can be made to be held in common for acoustic and optical metamaterials.

5.3   OPTICAL METAMATERIALS

The physics of optical metamaterials is similar to that of acoustic metamaterials. In both systems resonator units are scattered throughout a background medium to form a metamaterial composite [7, 11, 42, 43]. The new characteristics of the metamaterials arise in a large part from the response of the resonator units to an external stimulus applied to the metamaterial. In acoustic metamaterials the response follows from an acoustic stimulus whereas in optical metamaterials the response follows from optical stimuli. In the case of either of these types of metamaterials the medium can be engineered to exhibit, respectively, acoustic or optical responses outside the range of those found in naturally occurring materials.

As an example, in acoustic systems the mass-shell-enclosed-mass resonator unit was encountered [12, 13]. It is designed to display a characteristic resonant frequency response corresponding to the relative motion of the mass shell with its enclosed mass near and at resonance. Only at frequencies near the resonance frequency is the negative mass effect exhibited by the metamaterial designed with such resonant features. The negative mass response comes when an appropriate external stimulus is applied to the media and embedded resonator units of the metamaterial. Consequently, the array of resonator units scattered through the background medium is an essential artificial component in formulating negative mass acoustic metamaterials.

Similarly, a resonator component for optical systems is an electromagnetic split ring resonator unit [11, 16, 25–27, 43]. It is developed as an engineered inclusion in a dielectric medium, enabling the design of systems exhibiting new electromagnetic properties. In this scheme, an array of split ring resonators embedded within a dielectric medium alters the optical response of the resulting metamaterial composite to an external electromagnetic stimulus. Each split ring resonator is constructed so as to offer a resonant electromagnetic response similar to that of a basic LC resonator circuit (such an LC circuit is illustrated in Fig. 5.5a), and the focus in the following is on studying the response of a split ring resonator and of an array of such resonator units to an external applied electromagnetic stimulus [11, 43].

The LC resonator is the most basic type of resonator unit studied in electrodynamics. It is characterized by a natural oscillatory electrodynamic display in time, with a resonant frequency given by [11]

$$\omega_0 = \sqrt{\frac{1}{LC}}. \tag{5.36}$$

At the resonant frequency the circuit exhibits a harmonic motion of charges and currents between the inductor and capacitor. The oscillator behavior of the unit is particularly of interest in the design of metamaterials when it is driven by an externally applied electromagnetic field with a magnetic field component directed perpendicular to the plane of the circuit. Under such conditions, important types of optical metamaterials can be created by embedding an array of LC resonators in a background dielectric medium. When properly designed such a material provides important new responses to external electromagnetic stimuli. In the following, a discussion will be given of the realization of LC circuits as embedded features in the dielectric medium of a metamaterials and of the coupling of the LC circuits to an external field [11].

The realization of features with behaviors characterized by the LC circuit in Fig. 5.5a is readily fabricated. C-shaped metal inclusions of the form in Fig. 5.5b are found to be characterized as LC circuits such as that in Fig. 5.5a. In this characterization, the C-shaped structure is a basic split ring resonator, though in practice more complex structures are needed to provide an optimal response of the LC resonator. (For example, usually a double C-structure is employed in which the resonator is formed from two coplanar C's, one placed inside and oppositely faced

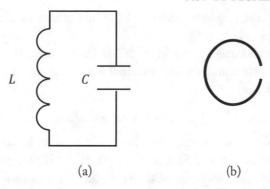

(a) (b)

Figure 5.5: Schematics of: (a) a basic LC resonator circuit and (b) the resonator circuit realized as a split ring resonator.

from the other.) Here only the basic idea behind the split ring resonator circuits will be presented, and the reader is referred to the literature for more details on specific designs needed to meet the requirements of technology [11].

In the split ring resonator of Fig. 5.5b the gap in the C-forms the capacitor represented diagrammatically in Fig. 5.5a. The inductor part of the circuit, represented in Fig. 5.5a, is formed by the partial ring of the C in Fig. 5.5b. The resulting basic resonator unit with resonance frequency ω_0 can be made to interact with the world outside of it through the application of Faraday's law. Both of the simple LC resonator units in Fig. 5.5 can be coupled by Faraday's law to an applied time-dependent magnetic field which is directed perpendicular to the page. With this type of coupling the circuits represented in Fig. 5.5 each become driven harmonic oscillators where the external field is the driving mechanism.

The current flowing in the C-inductor ring of the split ring resonator gives rise to a magnetic moment of the ring, and this shows up as a magnetic response of the metamaterials composed from the resonator units. For a current, $I(t)$, flowing in a ring enclosing an area, A, a magnetic moment is generated at the ring which has a magnitude given by [11, 43]

$$\mu \propto AI. \tag{5.37}$$

Depending on the direction of the current in the loop of the ring, the magnetic moment of the resonator is a diamagnetic or paramagnetic response, but since the current depends on time the intensity of the response is time-dependent.

What is of interest for applications to technology, however, is the time averaged magnetic moment generated by the externally applied field. On average, for a time-dependent applied field the electromagnetic responses of the split ring resonators range from a paramagnetic to a diamagnetic response as the frequency of the externally applied radiation is tuned through the ω_0 resonance frequency. This is found from the result for the dipole moment of a single split ring by studying the average magnetic energy of the interaction of the external field with the moment of the split ring [43].

For frequencies in the neighborhood of the LC resonance, the average moment of the simple ring in Fig. 5.5b interacting with an external field of frequency ω is given as a function of frequency by the form [11]

$$\mu_{average} \propto -(\omega - \omega_0)\, B_0, \tag{5.38}$$

where $B_0 > 0$ is the amplitude of the magnetic induction perpendicular to the plane of the ring. Consequently, it is found in Eq. (5.38) that the sign of the dipole moment generated by an applied field changes at ω_0. The response of the ring goes from diamagnetic (induced magnetic moment opposite the applied field) to paramagnetic (induced moment parallel to the field) as the frequency of the applied field passes through ω_0. At $\omega = \omega_0$, however, there is no net interaction of the applied field with the split ring resonator unit.

The nature of the frequency dependence of the magnetic moment in Eq. (5.38) provides for an enhanced magnetic response in bands of frequency near the resonance frequency of the split ring resonators. This is found from a detailed consideration of all the split ring resonators embedded in the dielectric of the metamaterial and from a treatment of the resonator responses at all frequencies. Consequently, on time-average the array of split rings in the metamaterial exhibits either a paramagnetic or diamagnetic response, and this response can be greatly enhanced over that found in naturally occurring materials. As shall be discussed later, this enhanced response is particularly important in the case of a diamagnetic response.

The diamagnetic response of the split ring resonator is similar to that found from the orbital motion of electrons in atoms, molecules, and conducting materials. However, since the length scales of the split ring resonators are different from those of atoms and molecules, the frequency responses of the split ring resonators and the atomic and molecular systems can be quite different. As shall be seen later, the split ring resonator units allow for the design of new materials (i.e., metamaterials) with diamagnetic responses that cannot be achieved in naturally occurring atomic and molecular media.

In the actual design of optical metamaterials not only do the split ring resonators interact with the external fields [11, 42, 43], but they also generate electromagnetic fields of their own that are seen by other split ring resonators of the system. Consequently, a complex system of split ring resonators interacting with one another and the externally applied fields is created. The metamaterials engineered in this way involve several types of interactions, making the study of their properties a complicated many body problem. In this regard, the detailed treatment of the problem of metamaterial response to applied fields usually involves the use of computer simulations. For the details of such treatments the reader is referred to the literature [11, 42, 43].

The observed enhancement of the magnetic response from many-body arrays of split ring resonators has an important application in the design of metamaterials with tailored magnetic responses not found in naturally occurring media. In this regard, computer-generated modeled metamaterials have been a focus of a number of material science applications. Later it will be seen that the region of frequencies of enhanced diamagnetic response are of particular use in technology, leading to metamaterials exhibiting negative refractive index, super lensing, and cloaking behaviors [11, 43].

To give an idea of the nature of such many-body considerations, however, a simplified one-dimensional model will be studied here. In particular, a one-dimensional array of weakly coupled split ring resonators is treated [18, 19]. This involves a chain of mutually inductive coupled split ring resonators with weak nearest neighbor inductive couplings. The electrodynamics of such a system is obtained as a set of magneto-inductive modal solutions of the system.

In the following, a brief discussion of the magneto-inductive modes of an array of split ring resonators is given. Afterward an introduction to negative refractive index optical materials is presented.

5.3.1   MAGNETO-INDUCTIVE WAVES

In this section a basic model of magneto-inductive waves, designed as a periodic array of split ring resonators, is presented [18, 19]. For simplicity, the model is limited to a one-dimensional chain, though the extension of the considerations to higher dimensions is straightforward.

A schematic of the chain is given in Fig. 5.6, showing an array of split ring resonators with each resonator represented as an LC circuit element. The neighboring LC circuits along the chain are coupled together by mutual inductance, and for simplicity only nearest neighbor circuits are considered to be inductively coupled to one another. Farther than nearest neighbor interactions are ignored as are radiative losses from the resonators. This constitutes the weak coupling limit of the chain and the case in which each of the LC circuits is a poor radiator of the modes propagating along the chain.

The energy of the infinite chain of inductively coupled LC circuits is given by [19]

$$E = \sum_{n=-\infty}^{\infty} \left[\frac{1}{2} L I_n^2 + M I_n I_{n+1} + \frac{1}{2} \frac{1}{C} Q_n^2 \right], \tag{5.39}$$

where L is the self-inductance of the split ring resonator ring, M is the nearest neighbor mutual inductance between the rings, and C is the capacitance of the split in the split rings realizing the individual LC circuit units. Ignoring radiative effects and the losses due to ring resistance, the system behaves as a conservative system. Consequently, in formulating the dynamics of the chain the energy in Eq. (5.39) is a constant of the motion.

Taking the time derivative of the energy, the equations of motion of the rings are obtained as [19]

$$L \frac{d^2 Q_n}{dt^2} + M \left(\frac{d^2 Q_{n+1}}{dt^2} + \frac{d^2 Q_{n-1}}{dt^2} \right) + \frac{1}{C} Q_n = 0. \tag{5.40}$$

Here the first and last terms on the left of the equation form the equation of motion for a single resonator ring and the middle term gives a coupling between the various LC resonators along the chain.

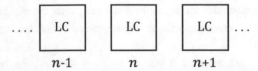

Figure 5.6: One-dimensional chain of LC circuits which are coupled by mutual inductance. Each resonator is denoted by a square labeled LC and represents one of the units in Fig. 5.5a. The center between nearest neighboring LC resonators is a. The resonators are labeled consecutively by integers ..., n-1, n, n+1, ...

The modal solutions of the set of linear difference equations in Eq. (5.40) are of the form of planewaves. Assuming solutions written as

$$Q_n = Q e^{i(kna-\omega t)}, \qquad (5.41)$$

where a is the nearest neighbor separation along the chain, gives upon substitution into Eq. (5.40), a dispersion relation for the modal excitations. In this way, the form of the dispersion relation is found to be

$$\omega^2 = \frac{1}{LC} \frac{1}{1 + 2\frac{M}{L}\cos ka}. \qquad (5.42a)$$

Here the values of the wavevectors, k, are a discrete set, which for periodic boundary conditions on a chain of $2N \to \infty$ LC circuits are given by

$$k = \frac{n\pi}{Na} \qquad (5.42b)$$

for n an integer.

Notice that in the limit that $M = 0$, when the LC oscillators of the chain are uncoupled, the dispersion relation in Eqs. (5.42) reduces to the resonant frequency of each of the set of isolated split ring resonators. For non-zero couplings, M may take a positive or negative value depending on the orientation of the plane of the rings relative to the axis of the chain. In this case, a weak dependence on k is found, and the frequency is increased or decreased from that of the modes of an isolated ring.

With the addition of a coupling to an external field the equations of motion take the form [19]

$$L\frac{d^2 Q_n}{dt^2} + M\left(\frac{d^2 Q_{n+1}}{dt^2} + \frac{d^2 Q_{n-1}}{dt^2}\right) + \frac{1}{C}Q_n = B_0 \sin\left(qna - \omega_{applied}t\right), \qquad (5.43)$$

where q is the wavenumber of the applied field, $\omega_{applied}$ is the frequency of the applied field, and B_0 is the coupling between the field and resonators. In the later discussions of metamaterials formed as arrays of coupled LC resonators the focus will be on the limit $q \approx 0$ of Eq. (5.43). In this limit the metamaterial behaves as a homogeneous medium in its response to the external field.

The behavior of the chain of coupled split ring oscillators studied here is that of a system of harmonic oscillators that are weakly coupled to one another. In this regard, the system is readily generalized to handle higher dimensional arrays and farther than nearest neighbor couplings. For the case in which the LC circuits are strongly coupled and for more complex designs of the split ring resonator units, the study of the properties of the system require computer simulation techniques. The reader is referred to the literature for these more advanced treatments [11, 43].

5.4   OPTICAL MATERIALS WITH NEGATIVE INDEX OF REFRACTION

The importance of optical metamaterials, composed as arrays of split ring resonators [11, 43], comes from their application in the design of optical media displaying a negative index of refraction. Similarly, the so-called double-negative acoustic metamaterials allow for media which can display a negative index of refraction [11, 25–27] in the propagation of sound. Both of these types of metamaterials exhibit novel refractive and propagation properties due to the resonator units that they incorporate in their structures. The metamaterials formulated from these resonator units offer possible applications in optical and acoustic superlensing effects, cloaking devices, and other phenomena requiring an enhanced range of steering of waves in optical or acoustic medium and at the interfaces of different optical and acoustic media.

As with acoustic metamaterials, it is the incorporation of resonant units into the medium which gives rise to the important new response to applied stimuli displayed by the resulting optical metamaterial [7]. For example, in terms of their propagation characteristics, the split ring optical metamaterials are analogous to the acoustic metamaterials of simultaneous negative effect mass and negative effective spring constants. These are the so-called double-negative acoustic metamaterials [14]. However, whereas the acoustic metamaterial requires two different type of resonators to display the appropriate propagation characteristics, the optical metamaterials only require a single type of split ring resonator.

As shall be seen both types of acoustic and optical metamaterials can be designed to exhibit frequency regions in which the energy propagation in the medium is in a direction opposite to that of the phase velocity wavevector [11, 43]. This particular type of propagation is the focus of important new properties in both optical and acoustic systems. In particular, it is found that materials with the characteristics of oppositely directed phase and group velocities are characterized as media which display a negative index of refraction.

Earlier in Eqs. (5.30) and Eq. (5.33)–(5.35) conditions were determined under which double-negative acoustic metamaterials exhibit propagating waves with oppositely directed group and phase velocities. There it was shown that both the effective Young's modulus and effective mass of the medium should exhibit the atypical properties of being negative and that these conditions were met only over a finite interval of frequencies [14, 15]. A similar situation shall now be shown to occur in optical metamaterials in which both the effective permeability and permittivity of the optical medium are negative [27]. These are the double-negative optical metamaterials.

5.4.1 DOUBLE NEGATIVE OPTICAL METAMATERIALS

Similar properties to those studied in acoustic media relating the phase, group, and energy transport velocities to one another where found in Chapter 2 for the propagation of planewaves in a homogeneous optical medium. Whereas in acoustic media the Young's modulus and mass density determined the nature of the propagation of planewaves in the system, in Section 2 of Chapter 2 it was shown that in optical media the permittivity and permeability set the nature of the propagation of electromagnetic planewaves.

In particular, for the propagation of a plane wave in an optical medium characterized by the permittivity, ε, and permeability, μ, it was found in Eq. (2.68) of Chapter 2 that for planewave propagation in the $\vec{k}$-direction the phase velocity relationship was given by

$$\vec{v}_p = \frac{\omega}{k}\hat{k} = \frac{c}{\sqrt{\mu\varepsilon}}\hat{k} \tag{5.44}$$

and the Poynting vector for energy flow is given by

$$\vec{S} = \frac{c}{8\pi}\frac{\sqrt{\mu\varepsilon}}{\mu}\left|\vec{E}_0\right|^2\hat{k}. \tag{5.45}$$

From these two equations it is seen that for $\mu, \varepsilon > 0$ propagating solutions exist for which the energy flow is in the same direction as the phase velocity. In the case that $\mu, \varepsilon < 0$, however, propagating solutions again exist but with the energy flow in the opposite direction to that of the phase velocity [11].

As with the acoustic media, the cases in which one of μ or ε is positive and the other is negative results in a non-propagating solution. These solutions exhibit an exponential increase or decrease and shall not be further considered.

In both the optical and acoustic media, the case in which the energy flow is opposite the phase velocity and wavevector of planewave solutions results in unusual refractive properties. In optical metamaterials [11], this is seen in the cases in which a planewave passes between two media (labeled 1 and (2) in which $\mu_1, \varepsilon_1 > 0$ and $\mu_2, \varepsilon_2 < 0$ or in which $\mu_2, \varepsilon_2 > 0$ and $\mu_1, \varepsilon_1 < 0$. For acoustic media, however, the new passage effects are between a medium with positive mass and Young's modulus and one with negative mass and Young's modulus. The details of these refractive properties will be discussed later for cases involving both optical and acoustic systems, but first the design of metamaterials with $\mu, \varepsilon < 0$ will be considered.

5.4.2   DESIGN OF OPTICAL METAMATERIALS

It is unfortunate that in naturally occurring optical media there is an absence of frequency bands in which $\mu, \varepsilon < 0$. This is a fundamental limitation and is primarily due to restrictions on the presence of the needed diamagnetic properties in crystalline, amorphous, and liquid materials found in nature. Media with $\varepsilon < 0$, on the other hand, are less difficult to locate so pose less of a problem in the design of $\mu, \varepsilon < 0$ materials. While various types of natural occurring and composite materials can be readily formulated which exhibit negative permittivities, materials with a suitable frequency range of diamagnetic permeabilities are less easily generated. Nevertheless, the problem of finding materials with the desired diamagnetic response is solved by the engineering of metamaterials. These can be designed to exhibit the appropriate diamagnetic frequency responses which would otherwise be absent and is a focus of the study of metamaterials designed to include split ring resonator components [11, 43].

A metamaterial formed as a periodic array of split ring resonators can be engineered to exhibit a diamagnetic response at a characteristic band of frequencies. This is done by designing appropriate split ring resonators and correctly arranging them in a periodic array within the background dielectric medium of the metamaterial. The split ring resonators are designed to exhibit a resonant frequency ω_0 and give an appropriate diamagnetic response to an excitation at a nearby frequency ω.

To function successfully, a number of conditions must be met in the design of the optical metamaterial for it to perform its function as a medium in which the group and phase velocities are opposite one another. In this regard, the wavelength of the excitation in the media must be large compared to the size and spacing of the embedded resonators so that the metamaterial appears to be homogeneous. In addition, for the correct permittivity response, the background medium into which the split ring resonators are embedded must have negative effective permittivity and its composite structure must also involve length scales much less than the wavelength of the excitation.

In the long wavelength limit the permeability, $\vec{\mu}$, of a metamaterial formed as an array of split ring resonators is expressed in terms of the magnetization, $\vec{M}$, of the array. The magnetization is in turn expressed in terms of the average magnetic moment of a resonator unit $\vec{\mu}_{res}$ and the density of resonator dipoles. It is found then that

$$\vec{M} = N \vec{\mu}_{res} \tag{5.46}$$

where N is the density of the dipoles per volume. From the constitutive field relation $\vec{H} = \vec{B} - 4\pi \vec{M}$ it follows for a simple medium in which all three of $\vec{H}, \vec{B}, \vec{M}$ are parallel and linearly related to one another that the metamaterial permeability is obtained as

$$\frac{B}{H} = \mu = 1 + 4\pi \frac{M}{H}, \tag{5.47a}$$

where from our previous considerations of the resonator response it follows that

$$\frac{M}{H} \propto -N\left(\omega - \omega_0\right). \tag{5.47b}$$

It is seen, from the form of Eqs. (5.47), that for appropriately chosen parameters a negative permeability can be generated near the resonant frequency. Consequently, a frequency band of diamagnetic response for the metamaterials occurs near ω_0.

The above simple treatment does not include crystal field effects which must be include for a strongly coupled array of interacting split ring resonators [11, 43]. However, the qualitative results for the diamagnetic properties of the metamaterial from such a more complete treatment are similar to those in Eqs. (5.46) and (5.47) for the weakly interacting system. In particular, it is found that there exists a region near the LC resonance frequency over which the array of resonators can be composed so as to exhibit an enhanced diamagnetic response to an applied electromagnetic field.

When the diamagnetic response of the resonator array is combined with a negative permittivity response from the background medium into which the resonator array is embedded, the resulting metamaterial will display a negative refractive index. This enhancement of the refractive features of optical media allows for new mechanisms in the optical manipulation of light. In the following, some of the physical properties that are found in the refraction of waves as they pass between positive and negative indexed media will be discussed. In this treatment it is taken for granted that homogeneous and isotropic three-dimensional metamaterials exist that are characterized by a negative index of refraction.

The types of three-dimensional homogeneous optical metamaterials discussed earlier have been formulated based on three-dimensional periodic arrays of split ring resonators. They are composed of arrays of split ring resonators formed into three different infinite sets. One set of resonators is a periodic array with the planes of the resonators being parallel to the $x-y$ plane. A second set of resonators forms a periodic array with the planes of the resonators being parallel to the $x-z$ plane. The final set of resonators is a periodic array with the resonator planes being parallel to the $y-z$ plane. For waves with wavelengths much greater that the resonator units and the nearest neighbor separations the arrangement responds as the homogeneous isotropic medium in three-dimensions.

In acoustic systems [7], similar types of ideas can be used to design three-dimensional acoustic metamaterials with homogeneous, isotropic, responses to long wavelength applied acoustic stimuli. The design of three-dimensional media with negative refractive indexes involves an embedding of various types of acoustic resonator arrays into an elastic background media. Some of the details of such arrays will be reviewed later, but in the following it will be assumed that homogeneous isotropic acoustic media with negative index exist. For a more complete treatment of the details of the design of both the optical and acoustic metamaterials the reader is referred to the literature.

5.5 REFRACTION AT AN INTERFACE

In the following an example is given of refraction of waves between two different media [11]. The focus is on a comparison of the refraction of waves passing between two positive indexed media with that of refracted waves passing from a positive indexed medium into a negative indexed medium. In both treatments it is meant to study a general model which could apply to either an optical or an acoustic system. Consequently, the parameters occurring in the models can easily be redefined to give representations of the refraction in either optical or acoustic systems.

5.5.1 GOING FROM POSITIVE TO POSITIVE INDEXED MEDIA

Consider an interface between two homogeneous isotropic media [11]. The two media are separated by a planar interface at $z = 0$ and are characterized by index of refraction $n_1 > 0$ in the region above the interface and $n_2 > 0$ below the interface. (See Fig. 5.7a for a schematic representation.) Here the index of refraction is defined in terms of the ratios of the phase velocities between different optical or acoustic medium, with its sign related to the parallel or antiparallel nature of the phase and group velocities in the medium it represents.

An incident planewave propagating in n_1 toward the planar interface is refracted into a planewave in n_2 propagating away from the interface and is also reflected into a planewave in n_1 propagating away from the interface. The incident wave is characterized by the planewave form

$$e^{i(k_x x - k_z z - \omega t)} \tag{5.48a}$$

for $k_x, k_z > 0$. Since $n_1 > 0$ the energy flowing in the planewave is parallel to the wavevector $(k_x, -k_z)$ and moves with the wave toward the interface.

Due to the translational symmetry of the system parallel to the interface both the reflected and transmitted waves must have the same wavevector components parallel to the interface. (In the language of group theory this is a consequence that the wavefunctions of the system must represent the infinite Abelian group of translational symmetries along the interface.) Consequently, the reflected wave is of the planewave form

$$e^{i(k_x x + k_z z - \omega t)} \tag{5.48b}$$

representing a phase velocity and energy flow away from the interface. In addition, the transmitted wave is of the planewave form

$$e^{i(k_x x - q_z z - \omega t)}, \tag{5.48c}$$

where $q_z > 0$ represents a phase velocity and energy flow away from the interface. (Notice that because the indices $n_1 \neq n_2$ the wavevector components $k_z \neq q_z$.)

In all three wavefunction forms in Eqs. (5.48) the energy flow is parallel to the wavevectors of the waves. As a result of this relationship between the energy flow and the wavevectors, the

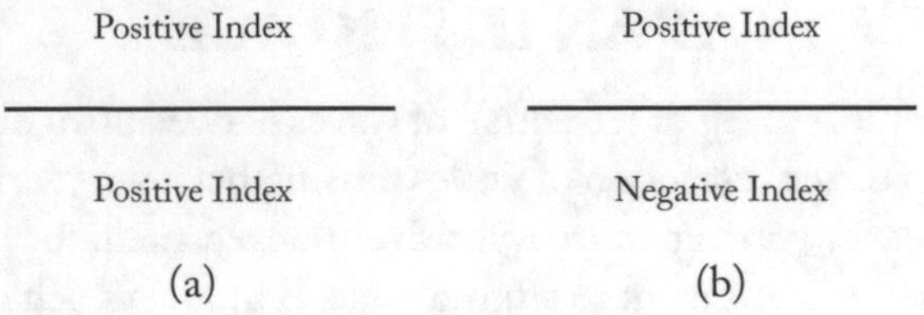

Positive Index   Positive Index

_____   _____

Positive Index   Negative Index

(a)   (b)

Figure 5.7: Schematic of a planar interface at $z = 0$ separating two different media, where the z-axis is vertical. In (a) the separation is between two media of $n_1, n_2 > 0$ and in (b) $n_1 > 0$ above the interface and $n_2 < 0$ below the interface. In the figure the horizontal line is the x-axis.

energy flow of the incident wave is toward the interface while those of the reflected and refracted waves are away from the interface.

Notice that the incident, reflected, and transmitted components of the wavefunction in Eqs.(5.48) have the same x-dependence given by $e^{ik_x x}$. Consequently, under the spatial translation $x \rightarrow x + T$ each of these forms picks up the same phase, $e^{ik_x T}$. The total wavefunction composed from the Eqs.(5.48) components is then one of the irreducible representations of the translational symmetry group along the interface.

From Eqs.(5.48) it is also found that for a wave incident in the second quadrant, the reflected wave is in the first quadrant, and the refracted wave is in the fourth quadrant. This is illustrated schematically in Fig. 5.8a. With refraction involving a passage between two positive index media, the three planewaves propagating in the system (i.e., the incident, reflected, and refracted waves) are fundamentally restricted to these quadrants and their passages. This is the case for double positive media (i.e., positive permittivities and permeabilities in optical media or positive masses and positive Young's modulus in acoustic media) and represents a fundamental limitation in the refraction properties of naturally occurring optical media.

The refractive properties of a wave incident from a positive index medium into a negative index medium are, however, quite different. These properties are now considered.

5.5.2   GOING FROM POSITIVE TO NEGATIVE INDEXED MEDIA

Consider the interface in Fig. 5.7b between two homogeneous isotropic media [11]. The two media are separated at $z = 0$ by a planar interface and are characterized by index of refraction $n_1 > 0$ in the region above the interface and $n_2 < 0$ below the interface. Unlike the system of Fig. 5.7a the medium below the interface is now of negative refractive index.

An incident planewave propagating in $n_1 > 0$ toward the planar interface is refracted into a planewave in $n_2 < 0$ with an energy that propagates away from the interface and is also reflected into a planewave in $n_1 > 0$ propagating away from the interface. The the planewave form of the incident wave is

$$e^{i(k_x x - k_z z - \omega t)} \tag{5.49a}$$

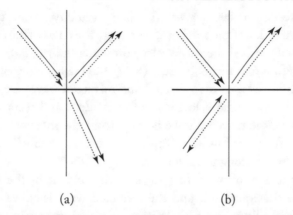

(a)   (b)

Figure 5.8: Schematics of the reflection and refraction of an incident planewave on a planar interface between two dielectric media. In (a) is an interface between two different positive indexed media, and in (b) is an interface between a positive index medium in the upper-half plane and a negative index medium in the lower-half plane. The solid arrows represent the directions of the wavevectors of the waves and the dashed arrows represent the directions of the energy flows in the waves. The incident waves are incident in the second quadrant and the reflected waves travel in the first quadrant.

for $k_x, k_z > 0$, representing an energy flow in the $n_1 > 0$ medium that is parallel to the wavevector $(k_x, -k_z)$. Both the wavevector and the energy flow are then directed toward the interface. In addition, the reflected wave in the $n_1 > 0$ medium is of the planewave form

$$e^{i(k_x x + k_z z - \omega t)}, \tag{5.49b}$$

and represents a phase velocity and energy flow away from the interface. These forms and their energy flows are standard in the treatment of media with positive index of refraction.

Care, however, must be taken when considering the transmitted wave in the negative index medium as it represents a situation not found in common classical optical and acoustic systems. In particular, the direction of the energy flow in the negative index medium is opposite the phase velocity which is in the direction of the wavevector of the transmitted planewave. Consequently, the wavevector in the $n_2 < 0$ medium must point toward the interface. In addition, due to the translational symmetry along the interface of the two media, the x-component of the wavevector must agree with those in the planewave forms in Eqs. (5.49a) and (5.49b).

With these restrictions the transmitted wave must have the planewave form

$$e^{i(k_x x + q_z z - \omega t)} \tag{5.49c}$$

where $q_z > 0$. This represents a plane wave with a phase velocity toward the interface, but, due to the negative refractive index of the medium, an energy flow away from the interface. (Notice that because the indices $n_1 \neq |n_2|$ the wavevector components $k_z \neq q_z$.)

In the three wavefunction forms in Eqs. (5.49) the energy flow of the incident wave is toward the interface while those of the reflected and refracted waves are away from the interface. On the other hand, the phase velocities of the incident and refracted waves are toward the interface while that of the reflected wave is away from the interface. In addition, the total wavefunction composed from the Eqs. (5.49) components is an irreducible representation of the translational symmetry group along the interface.

From Eqs.(5.49) it is also found that for a wave incident in the second quadrant, the reflected wave is in the first quadrant, and the refracted wave is now in the third quadrant. This is illustrated schematically in Fig. 5.8b. With refraction involving a passage from a positive index medium to a negative index medium, the three planewaves propagating in the system (i.e., the incident, reflected, and refracted waves) are fundamentally restricted to these quadrants and their passages.

5.5.3   SNELL'S LAW

As a summary of the above discussions, it is found from a detailed consideration of the boundary conditions that the refraction of light or acoustic radiation is described by Snell's Law. This relates the angle of incidence of the incident wave to the angle of refraction of the refracted wave for transmission through a planar interface separating two different media.

Considering the geometry in Fig. 5.9, the upper-half plane contains the n_1 medium and the lower-half plane contains the n_2 medium. The angle of incidence θ_i is measure from the vertical axis in the n_1 medium, while the angle of refraction θ_r is measure from the vertical axis in the n_2 medium. The positive sense of both angles is in the anti-clockwise sense whereas the negative sense of both angles is in the clockwise sense.

With these definitions and the geometry of Fig. 5.9, Snell's law becomes

$$n_1\sin \theta_i = n_2\sin \theta_r. \tag{5.50}$$

This form of Snell's law can now be used to study the ray optics of the variety of motions of light considered in geometric optics. In addition, the reflected ray from the interface is not indicated in Fig. 5.9 but is found in all cases to obey the standard relation that the angle of incidence is equal to the angle of reflection. The law of refection is, consequently, independent of the sign of the index of refraction of the media.

5.6   SOME APPLICATIONS OF NEGATIVE INDEX MATERIALS

One of the simplest examples of a new optical effect arising from the negative indexed materials involves the passage of light through an infinite slab of negative index medium [11, 26]. The slab

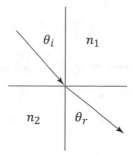

Figure 5.9: Refraction at a planar interface (horizontal line) between medium n_1 and medium n_2. The angle of incidence θ_i is measure from the vertical axis in the n_1 medium, while the angle of refraction θ_r is measure from the vertical axis in the n_2 medium. The positive sense of both angles is in the anti-clockwise sense whereas the negative sense of both angles is in the clockwise sense.

is surrounded by vacuum and a planewave is incident on the slab from vacuum, being transmitted to the other side of the slab. The increased bending or refraction of light at the slab surfaces of the negative index slab over that at a slab of positive index medium shows up as a difference in the lateral shifting of the rays of light as they pass through the slab.

This effect is illustrated in the two figures describing such passage in Fig. 5.10. One figure describes the passage of light through a positive index slab, and the other describes the passage through a negative index slab [11, 26]. Note that in these figures the reflected waves at the surfaces are ignored, and the slab surfaces are parallel planes.

In Fig. 5.10a the passage of a ray of light through a slab of positive index medium is illustrated. The refraction at the top surface of the slab shifts the trajectory of the incident ray (incident in the second quadrant) into a refracted ray in the slab (refracted into the fourth quadrant). At the lower surface of the slab the incident ray (incident in the second quadrant) is again refracted, passing into a refracted ray (refracted into the fourth quadrant) moving parallel to the original incident ray above the slab. The final refracted ray is moved to a trajectory which is shifted laterally to the right of the original incident ray.

Next consider the same passage for the case in which the slab is composed of a medium with a negative index. In Fig. 5.10b the passage of a ray of light through a slab of negative index medium is illustrated. The refraction at the top surface of the slab shifts the trajectory of an incident ray (incident in the second quadrant) into its refracted ray in the slab (refracted now into the third quadrant). At the lower surface of the slab the ray is again refracted into a refracted ray moving parallel to the original incident ray above the slab. At the lower surface the refraction sends the incident ray in the first quadrant to a refracted ray in the fourth quadrant. The final refracted ray moves to a trajectory which is shifted laterally in this case to the left of the original incident ray.

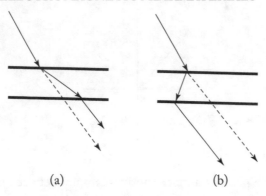

$$(a) \qquad\qquad\qquad (b)$$

Figure 5.10: Schematics of: (a) passage of a ray of light through a slab of positive index medium and (b) passage of a ray of light through a slab of negative index medium. The solid arrowed lines are the passage of a ray of light with the arrows indicating the direction of the motion of the light energy. The dotted arrowed lines indicate the passage of the incident ray in the absence of the slab. The reflected waves in the system have been ignored.

An addition property of the slab of negative index medium is that it can be designed to focus light. This is an unusual property arising from negative index media as in positive index media a focusing lens requires curved surfaces.

To see how an infinite slab geometry can act as a focusing lens, consider a slab formed of a negative index material with an index $n_s = -1$. The slab has parallel plane surfaces and is surrounded by vacuum. A schematic of the lens including an object-image ray diagram is shown in Fig. 5.11.

First consider the refraction at the top surface in Fig. 5.11. For the Snell's law geometry in Fig. 5.9 applied to the upper surface in Fig. 5.11, it follows from Eq. (5.50) that the incident and refracted rays are related by

$$\theta_s = -\theta_i, \tag{5.51a}$$

where θ_i is the angle of incidence of the incident ray and θ_s is the angle of refraction in the slab. Similarly, at the lower surface in Fig. 5.11 it follows from Eq. (5.50) that

$$\theta_r = -\theta_s = \theta_i, \tag{5.51b}$$

where θ_s is the angle of incidence at the lower surface of the incident ray in the slab and θ_r is the angle of refraction in the vacuum region below the slab. Notice that the rays above and below the slab travel parallel to one another.

By arranging the slab thickness and position of the object an image of the object is formed projected by the slab [26]. This is illustrated in Fig. 5.11.

Object

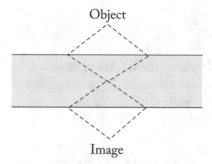

Image

Figure 5.11: Schematic of a rays of light passing through a perfect lens, going from an object above the slab to an image below the slab. The slab is an infinite slab of dielectric constant $n_s = -1$ surrounded by vacuum. The surfaces of the slab are infinite parallel planes and the rays are denoted by dashed lines.

In the figure, the slab is an infinite slab of dielectric constant $n_s = -1$ surrounded by vacuum, and the surfaces of the slab are infinite parallel planes. The rays are denoted by dashed lines and propagate from the object through the slab to form an image on the opposite side of the lens.

The lens is called a perfect lens as it creates an image of the object which also contains all of the modes of light, i.e., both the propagating and evanescent components. (For a proof of this the reader is referred to a more detailed treatment [11, 26].) In this regard, the image is a perfect reproduction of the object.

As a final point regarding the new range of refractive angles, it should be noted that this new feature in optical materials allows for an extended range in the channeling of light through space. For example, applications have been designed for the development of optical cloaking schemes in which light is made to travel around an object through an engineered medium composed of regions of both positive and negative index materials [11, 25, 26, 43]. In this way the light incident on the cloaked object travels around it so that the object within the cloak cannot be detected.

5.7   SOME FINAL CONSIDERATIONS AND RECENT DEVELOPMENTS

It should be noted that metamaterials based on the applications of principles of resonance have certain drawbacks [11, 43]. In this regard, associated with all resonances are losses and these losses should be inherent in all designs of resonant based metamaterials. Consequently, it is a further engineering problem to minimize losses occurring in the metamaterials.

To understand the nature of losses from resonance consider the resonance properties of a simple forced damped harmonic oscillator. The dynamics of the oscillator is described by the formula

$$m\frac{d^2x}{dt^2} + m\frac{1}{\tau}\frac{dx}{dt} + m\omega_0^2 x = F_0 e^{-i\omega t}. \tag{5.52}$$

In Eq. (5.52) the second term on the left is a phenomenological term describing an energy loss from the system, the right-hand term is the driving force which has the frequency ω, and the natural frequency of the harmonic oscillator is ω_0.

The solution of Eq. (5.52) is

$$x(\omega) = \frac{-F_0 e^{-i\omega t}}{m\left[\omega^2 - \omega_0^2 + i\omega\frac{1}{\tau}\right]}. \tag{5.53}$$

It is seen from Eq. (5.53) that for a physical solution, the imaginary part in the denominator in Eq. (5.53) must be nonzero. If this is not the case, then the resonance frequency of the system at $\omega = \omega_0$ is a singularity of the solution. Consequently, the losses represented by the imaginary parts in the denominator of Eq. (5.53) are most important in the system dynamics at the resonant frequency.

Resonance losses of this type are present in all resonator systems, and various ways of minimizing the effects of these losses have been proposed [7]. Such remedies are topics of a more advanced treatment, and the reader is referred to the literature for further discussions.

Hyperbolic Metamaterials

Hyperbolic metamaterials are another recently developed type of metamaterial [11, 45] that illustrate some interesting important new applications. They are engineered materials, designed to respond as homogeneous media with anisotropic dielectric properties. Light propagating in them exhibits unusual dispersive properties that are of technological interest due to their novel densities of states properties and their applications in far-field subwavelength imaging.

In a homogeneous media with a dielectric $\varepsilon > 0$, the dispersion relation of light is given by

$$\frac{k_x^2 + k_y^2 + k_z^2}{\varepsilon} = \frac{\omega^2}{c^2}. \tag{5.54}$$

A hyperbolic metamaterial, on the other hand, is a composite material designed to have an anisotropic dielectric tensor characterized by $\varepsilon_{xx} = \varepsilon_{yy}$, ε_{zz} yielding an anisotropic dispersion relation of the form

$$\frac{k_x^2 + k_y^2}{\varepsilon_{zz}} + \frac{k_z^2}{\varepsilon_{xx}} = \frac{\omega^2}{c^2} \tag{5.55}$$

where either $\varepsilon_{zz}, -\varepsilon_{xx} > 0$ or $-\varepsilon_{zz}, \varepsilon_{xx} > 0$. While at constant frequencies the wavevectors of the propagating modes of light in the isotropic medium lie on spheres, in the hyperbolic media the propagating modes have wavevectors restricted to hyperboloids. This has a consequence for the spatial information carried by light through the media [11, 43]. In imaging applications, the dispersion relation in Eq. (5.55) can resolve more spatial information than that in Eq. (5.54).

Hyperbolic media are designed as a layering of dielectric and metal planes or as an array of metal wires in a dielectric background. In the next chapter a treatment of properties of related layered media is made.

CHAPTER 6

Surfaces and Surface Waves

Other types of systems that can exhibit important new characteristics in the design of materials with enhanced physical properties are those involving interactions at surfaces [5, 46, 48–51]. As in the case of bulk photonic and phononic crystals and metamaterials, the properties of interest in surface systems arise from the nature of the excitations present in them. In this regard, earlier chapters have shown that in photonic and phononic crystals the presence of periodicity opens a series of frequency stop and pass bands introducing new physics as the basis of technological applications otherwise absent in homogeneous bulk materials. Similarly, it has been seen that the embedded resonant features present in the design of metamaterials introduce new frequency responses not otherwise available in a naturally occurring homogeneous medium. As shall now be found in this chapter, in the case of surfaces the presence of surface wave excitations facilitates the enhancement of physical properties that would otherwise be absent in bulk homogeneous materials. This is particularly evident in the difference between the properties of surfaces supporting and not supporting surface wave excitations and in the strong field enhancements possible for systems supporting surface wave excitations.

The interesting properties found at surfaces come from the presence of surface wave excitations with fields concentrated along the media interface [5, 46, 48–51]. Surface waves are modal solutions which are bound to and propagate along an interface between two media. They often require a rather specific set of conditions be present on the media and the parameters characterizing the interface geometry for their existence. In general, the amplitudes of the surface waves are large at the surface and decay exponentially to zero with increasing separation from the surface. Consequently, the energy within these excitations is often highly concentrated near the surface and moves parallel to the interface.

The surface waves have different propagation characteristics than bulk waves and exist as distinct excitations at different frequencies from the bulk excitations. Consequently, at a planar interface between two media bulk waves can pass from one bulk medium to the other independent of the presence of surface waves. In this regard, surface waves on the interface can only be excited on the interface by bulk waves that couple to the surface wave through geometric features on the interface. On planar interfaces, the surface waves are isolated from the bulk excitations, but the presences of roughness or surface bumps which destroy the translational symmetry of the surface creates a coupling to bulk excitations in the media. As a result of this coupling, surfaces waves can be excited by the incidence of bulk waves on the interface or they can decay into bulk waves radiating away from the interface.

The concentration of energy at the interface in surface wave excitations provides a useful mechanism employed as a basis of a number of technological applications. An important example is in the spectroscopic technique of surface enhanced Raman scattering (SERS) [5, 11, 52]. In this technique the excitation of a surface wave on an interface supporting absorbed molecules can be used to perform Raman spectroscopy on the molecules. The idea here is to perform spectroscopy by scattering surface waves on the absorbed molecules in the same manner as bulk waves are used to study molecular spectroscopy in the absence of surfaces. In this spectroscopy waves from the bulk are made to excite surface waves which then perform a spectroscopy on the absorbed molecules. The spectroscopy employing excited surface waves is enhanced over that in the absence of surface waves due to the increased intensity at the absorbed molecule of the surface waves over their bulk counterparts. In addition, it should be noted that the technique is not just restricted to molecules but can be used on a number of other types of features introduced to the surface and benefiting from the increased fields of surface waves on the interface.

Surface waves are also important in the explanation of a variety of enhanced transmission effects present at periodically patterned surfaces and in the design of near field microscopy schemes [11, 53]. For these applications devices based on slab geometries are often used to formulate technological applications of the ideas evolved from surface phenomena. In enhanced transmission phenomena light is incident on a periodically patterned slab from a direction which is normal incident on the plane of the surface which is also the plane of the periodicity of the pattern [11]. An enhancement in the light transmitted through the slab is found to arise from the excitation of and interaction between surface electromagnetic waves on the two surfaces of the transmission slab. Similarly, in near field microscopy [11, 54] the added information about a surface contained in the surface electromagnetic waves is extracted by the use of a near field probe. This allows for a subwavelength microscopy resolution of the interface features. Surface waves are at play in both of these new and developing technologies.

Another optical enhancement technique based on the excitation of surface electromagnetic waves is found in nonlinear optics [11, 22]. This involves the generation of second harmonics of radiation based on the application of a radiation at a fundamental frequency to a medium with a nonlinear polarization. Interactions of the fundamental with the nonlinearity of the medium generates a second harmonic component of radiation, and the intensity of the generated second harmonic depends, among other things, on the intensity of the applied field. In these nonlinear processes the field concentrating effects of surface electromagnetic waves is an aid in the generation process at surfaces.

In acoustics, surface waves [46] have similar field concentrating effects to those found in applications of electromagnetic surface waves. Such concentrating effects are well known from the study of earthquakes and in the generation of surface shape resonances. In addition, similar enhancement related phenomena to those discussed above for electromagnetic waves should be found in acoustic systems. In this regard, there is an analogy of electromagnetic and acoustic phenomena in the various applications of linear and nonlinear optical and acoustic technologies.

In the following chapter, first a discussion of common acoustic surface wave solutions is presented. These will include treatments of Rayleigh waves which are surface acoustic waves at a planar interface between two acoustic media and of Love waves which are guided acoustic wave solutions in an infinite slab of acoustic medium [5, 11, 46, 49, 50]. This is followed by the study of surface electromagnetic waves on a planar interface between two different dielectric media. As a conclusion, the bound wave modes on a layered dielectric media will be studied with a focus on the different types of solutions present in the layering and the field concentrating effects of these modes.

6.1   SURFACE ACOUSTIC WAVES

In the following, surface and guided wave acoustic excitations will be introduced [5, 46]. The solutions of the basic excitations at interfaces between elastic media have been important in a variety of crystalline surface studies which include treatments of surface scattering and the nature of bound resonant excitations associated with surface imperfections [46, 48–50]. The excitations are also of importance to seismology and geophysical studies. Consequently, the solutions presented here enter into many different areas of investigations. However, as a present focus, only the elementary solutions are presented, and the reader is referred to the extensive literature for studies of particular systems and their applications in technology [5, 46].

6.1.1   RAYLEIGH WAVES

The simplest type of surface acoustic wave is a Rayleigh wave [5, 46, 48]. It occurs as an elastic excitation moving along the interface between two different elastic media and is formed as a mixture of longitudinal and transverse displacement components. The resulting excitation has a large elastic displacement at the interface which decays in amplitude with increasing separation from the interface. In the following an outline of the Rayleigh wave solution at a planar interface between an elastic medium and vacuum is presented.

Consider a semi-infinite elastic medium in the region $z < 0$ and a region of vacuum for $z > 0$. Below the planar interface the elastic medium is isotropic homogeneous with a wave equation of the general form [46] (see the discussions leading to Eq. (2.29) in Chapter 2 for a development of this form):

$$\rho \ddot{\vec{u}} = (\lambda + 2\mu) \nabla (\nabla \cdot \vec{u}) - \mu \nabla \times (\nabla \times \vec{u}), \tag{6.1}$$

where $\vec{u}$ is the displacement vector, ρ is the mass density, and λ and μ are the Lame constants. The form of this equation was discussed in an earlier chapter in which a review of elasticity theory was given as well as a treatment of the Lame constants for a homogeneous isotropic medium. Above the interface is vacuum which is accounted for by the application of stress-free boundary conditions at the interface.

From the earlier discussions in Chapter 2 it was found that the displacement vector in Eq. (6.1) has the general form

$$\vec{u} = \nabla\varphi + \nabla \times \vec{\psi} \tag{6.2}$$

where the first term represents longitudinal motions and the second transverse motions [46]. Taking the divergence of Eq. (6.1) a Helmholtz equation is obtained with the form [46]

$$\frac{\partial^2 u_d}{\partial t^2} = \frac{\lambda + 2\mu}{\rho} \nabla^2 u_d, \tag{6.3}$$

where $u_d = \nabla \cdot \vec{u}$. Similarly, taking the curl of Eq. (6.1) generates a second Helmholtz equation given by

$$\frac{\partial^2 \vec{u}_c}{\partial t^2} = \frac{\mu}{\rho} \nabla^2 \vec{u}_c, \tag{6.4}$$

where $\vec{u}_c = \nabla \times \vec{u}$. From a study of these two equations the displacement fields in the isotropic homogeneous solid can be obtained through the application of Eqs. (6.1) and (6.2). This shall now be done for the case of planewave solutions.

A planewave solution to Eqs. (6.1)–(6.4) for the scalar components in Eq. (6.2) in the region $z < 0$ is obtained in the form

$$\varphi = -i a_1 \exp\left[q_1 z + i (kx - \omega t)\right]. \tag{6.5a}$$

This solution represents a wave propagating in the x-direction and, for $q_1 > 0$, of decaying amplitude with separation from the surface. The amplitude, $-i a_1$, of the wave is later determined by the boundary conditions. In addition, upon substitution of Eq. (6.5a) into Eq. (6.3) it follows that q_1, k, and ω are related to one another by

$$\frac{\lambda + 2\mu}{\rho} \left[q_1^2 - k^2\right] + \omega^2 = 0. \tag{6.5b}$$

A second solution to Eqs. (6.1)–(6.4) has the vector planewave form [46]

$$\vec{\psi} = -i a_2 \hat{j} \exp\left[q_2 z + i (kx - \omega t)\right]. \tag{6.6a}$$

This is a wave propagating in the x-direction with a transverse vector of amplitude $-i a_2$ in the y-direction, and as in the case of Eq. (6.5) $q_2 > 0$ is required to represent a wave decaying in amplitude with increasing separation from the interface. Substituting Eq. (6.6a) into Eq. (6.4), it is found that q_2, k, and ω are related by

$$\frac{\mu}{\rho} \left[q_2^2 - k^2\right] + \omega^2 = 0. \tag{6.6b}$$

In order that the solutions in Eqs. (6.5a) and (6.6a) represent components of a surface wave it is required that both the solutions for both q_1 and q_2 from Eqs. (6.5b) and (6.6b) are real and positive. Taking this into account, the scalar and vector solutions obtained in Eqs. (6.5) and (6.6) are now used to match the boundary conditions at the solid-vacuum interface. This will yield a set of equations with a solvability condition determining the Rayleigh wave dispersion relation.

The amplitudes a_1 and a_2 in Eqs. (6.5) and (6.6) are determined by the stress-free boundary conditions at the vacuum solid interface. In the process of applying the boundary conditions a set of two linear homogeneous algebraic equations for a_1 and a_2 are obtained, and the conditions that a solution of these exists determines the surface wave dispersion relation relating ω and k.

The stress-free boundary conditions at the $z = 0$ interface states that [46]

$$\sigma_{xz} = \mu \left(\frac{\partial u_x}{\partial z} + \frac{\partial u_z}{\partial x} \right) = 0 \tag{6.7a}$$

and that

$$\sigma_{zz} = \lambda \left(\frac{\partial u_x}{\partial x} + \frac{\partial u_z}{\partial z} \right) + 2\mu \frac{\partial u_z}{\partial z} = 0. \tag{6.7b}$$

Applying these conditions, from Eqs. (6.2), (6.5), (6.6), and (6.7) it then follows that the two equations for a_1 and a_2 are

$$2kq_1 a_1 - \left(k^2 + q_2^2 \right) a_2 = 0 \tag{6.8a}$$

and

$$\left[\lambda \left(k^2 - q_1^2 \right) - 2\mu q_1^2 \right] a_1 + \left[2\mu q_2 \right] k a_2 = 0. \tag{6.8b}$$

These form a linear homogeneous two-dimensional matrix eigenvalue problem.

The dispersion relation of the Rayleigh wave is obtained from the solvability condition on the two homogeneous equations in Eq. (6.8). In this way it is found that ω is obtained as a function of k from the solutions of

$$\left[1 - \frac{1}{2} \frac{\rho}{\mu} c_R^2 \right]^2 = \left[1 - \frac{\rho}{\lambda + 2\mu} c_R^2 \right]^{1/2} \left[1 - \frac{\rho}{\mu} c_R^2 \right]^{1/2}, \tag{6.9}$$

where $c_R = \frac{\omega}{k}$ is the Rayleigh wave phase velocity. Following some algebra, Eq. (6.9) is rewritten in the form of a cubic equation given by

$$\delta^3 - 8\delta^2 + 8 \left(3 - 2\tau \right) \delta - 16 \left(1 - \tau \right) = 0, \tag{6.10}$$

where $\delta = \frac{\rho}{\mu} c_R^2$ and $\tau = \frac{\mu}{\lambda + 2\mu}$. Given the solutions for c_R from Eq. (6.10) the amplitudes a_1 and a_2 are obtained directly from a substitution of c_R into Eq. (6.8).

As an example of a solution of Eq. (6.10), consider the case of the limiting form known as the Poisson limit. This is the simple case of a system in which $\lambda = \mu$. For such systems $\tau = \frac{1}{3}$, and a real solution of Eq. (6.10) is $\delta = 0.845$. In these types of systems, the of speed of the Rayleigh waves depend mainly on $\frac{\mu}{\rho}$ and is obtained as $c_R = \sqrt{\frac{\mu}{\rho}}\sqrt{0.845}$. Many other solutions exist for general μ, λ, ρ, but the Poisson limit is a common simple example studied in acoustics and often associated with problems in seismology.

The next solution of interest in acoustic studies is that of a guided or Love wave mode. For Love waves there are more than just a single planar interface in the system.

6.1.2   LOVE WAVES

A Love wave is another type of wave moving parallel to a planar interface [46, 51]. It is an acoustic wave that propagates along and is primarily concentrated within a slab layered on a substrate medium. The Love wave solutions in the slab are a form of guided wave, and outside the slab the displacement fields of the Love waves decay in amplitude with increasing separation from the slab. In their study many of the ideas of the Rayleigh wave solutions are employed in developing the theory of Love waves. An important difference between Love and Rayleigh waves, however, is that Love waves are solely transverse waves whereas Rayleigh waves are a mixture of longitudinal and transverse displacements. The transverse nature of the solutions is a simplification in the calculation of the properties of Love waves.

For the treatment of Love waves consider a layer or slab of isotropic acoustic medium characterized by the parameters ρ_1, μ_1 and layered on a substrate characterized by the parameters ρ_2, μ_2. The upper surface of the layer is at $z = 0$ and its interface with the substrate is at $z = -d$. The region $z < -d$ is filled by the substrate medium, and above the slab in the region $z > 0$ is a vacuum. In this geometry a shear acoustic wave known as a Love wave is concentrated within the layer and moves parallel to the slab surfaces. For the following discussions the Love wave is considered to propagate in the plane parallel to the interfaces and to move along the x-axis.

A Love wave composed of y-displacements for a wave moving in the x-direction within a medium described by the parameters ρ, μ has shear displacements which are solutions of Helmholtz equations of the form [46]

$$\frac{\partial^2 u_y}{\partial t^2} = \frac{\mu}{\rho}\nabla^2 u_y = c^2 \nabla^2 u_y \tag{6.11}$$

where c is the speed of sound. Here it is assumed that the wave is polarized along the y-axis which is perpendicular to the direction of propagation in the system and parallel to the interfaces in the system. Wave forms based on the solution of Eq. (6.11) exist in the two different media forming the slab and substrate and are matched up at the slab-substrate interface by acoustic boundary conditions. At the upper surface of the slab, on the interface with vacuum, the system solutions obey stress-free boundary conditions.

Within the slab the displacement of the elastic medium parallel to the y-axis is given by the form

$$u_y(x, z, t) = [A\exp(iq_1 z) + B\exp(-iq_1 z)]\exp[i(kx - \omega t)], \tag{6.12}$$

where $q_1^2 = k^2 \left[\frac{\rho_1 c_L^2}{\mu_1} - 1 \right] > 0$ and c_L is the speed of the Love wave. Consequently, the propagation within the slab is planewave propagation in both the x- and z-directions. Within the substrate, however, the displacement of the elastic medium parallel to the y-axis is given by the form

$$u_y(x, z, t) = C \exp(q_2 z) \exp[i(kx - \omega t)], \qquad (6.13)$$

where $q_2^2 = k^2 \left[1 - \frac{\rho_2 c_L^2}{\mu_2} \right] > 0$ and c_L is the speed of the Love wave. In this case the solution is seen to exhibit an exponential decay as one leaves the slab and travels further into the substrate medium. Above the slab is vacuum and no elastic wave is supported by the vacuum.

At the top of the substrate (i.e., at $z = 0$) the surface is free so that the shear stress at $z = 0$ is

$$\sigma_{zy} = 0. \qquad (6.14)$$

Applying this condition to Eq. (6.12) requires that $A = B$. On the other hand, at the interface between the slab and the substrate the displacement fields in the two media are continuous so that

$$2A \cos(q_1 d) - C \exp(-q_2 d) = 0 \qquad (6.15)$$

and the continuity of the shear stress ($\sigma_{yz} = 2\mu_i \frac{\partial u_y}{\partial z}$ for $i = 1, 2$) yields

$$2A\mu_1 q_1 \sin(q_1 d) - C\mu_2 q_2 \exp(-q_2 d) = 0. \qquad (6.16)$$

The condition of solution for the equations in Eqs. (6.15) and (6.16) is that

$$\tan(q_1 d) = \frac{\mu_2 q_2}{\mu_1 q_1}. \qquad (6.17)$$

This condition is in the form of a transcendental equation which is solved subject to the conditions $q_1^2 = k^2 \left[\frac{\rho_1 c_L^2}{\mu_1} - 1 \right] > 0$ and $q_2^2 = k^2 \left[1 - \frac{\rho_2 c_L^2}{\mu_2} \right] > 0$ where $c_L = \frac{\omega}{k}$ is the speed of the Love wave. A consequence of the conditions on q_1 and q_2 is that

$$\frac{\mu_1}{\rho_1} < c_L < \frac{\mu_2}{\rho_2} \qquad (6.18)$$

for Love waves to be formed in the system.

The dispersion relation of the Love waves is obtained from Eq. (6.17) under the conditions in Eq. (6.18). In this way, the transcendental equation can be rewritten as

$$\tan \left[kd \sqrt{\frac{c_L^2 \rho_1^2}{\mu_1^2} - 1} \right] = \frac{\mu_2}{\mu_1} \frac{\sqrt{1 - \frac{c_L^2 \rho_2^2}{\mu_2^2}}}{\sqrt{\frac{c_L^2 \rho_1^2}{\mu_1^2} - 1}}. \qquad (6.19)$$

Once the parameters of the elastic media are fixed, the solutions of Eq. (6.19) yield $\frac{c_L^2 \rho_1^2}{\mu_1^2}$ in terms of kd. In this way a set of multiple solutions of Eq. (6.19) are obtained for the dispersion relations of the various Love wave modes possible in the slab-substrate system.

The Rayleigh and Love waves are the two most important solutions for surface acoustic waves treated in phononic crystals and metamaterials. Similar types of surface excitations are found in the electrodynamic solutions of their photonic counterparts. These solutions are now discussed.

6.2   SURFACE ELECTROMAGNETIC WAVES

In this section surface and guided electromagnetic waves are introduced and studied in the context of simplified models possessing analytic solutions [5, 11, 30]. The electromagnetic results are found to have similar mathematical structures to those of their related Rayleigh and Love wave counterparts, treated earlier in the discussion of acoustic surfaces. Both the electromagnetic and acoustic theories are based on solutions of equations of the Helmholtz form, with the essential difference between the acoustic and electromagnetic treatments entering in the applications of the boundary conditions. Consequently, the theory of surface electromagnetic and guided waves exhibits a variety of similar technological applications to those arising from the theory of Rayleigh and Love waves and their applications.

Keeping this in mind (and since detail treatments of surface electromagnetic waves exist elsewhere [5, 11, 30]) only an outline of the electromagnetic surface wave solutions and their properties is presented here. An analysis of these solutions will indicate the differences between the acoustic and electromagnetic treatments. In addition to the electrodynamic results for surface waves, a further discussion of the guided electromagnetic modes in a layered medium composed of a finite number of layers on a substrate is presented. These last layered systems have been of recent interest in nanophotonics applications in the field of plasmonics, and some discussions of their important properties in nanoscience device applications are presented.

6.2.1   SURFACE ELECTROMAGNET WAVES ON A PLANAR INTERFACE BETWEEN TWO MEDIA

In the following, a treatment is given of basic surface electromagnetic modes bound to and traveling on the planar interface between vacuum and a dielectric medium. As was the case for the acoustic surface waves, the field intensities of the surface electromagnetic waves are most intense at the interface and decay with separation from the interface into the bulk. These surface waves represent a separate set of solutions which are independent of the electromagnetic modes in the bulk.

To understand the nature of the surface wave excitations, consider a planar interface located at $x = 0$ which separates two homogeneous regions with different dielectric properties. A commonly studied model is treated here involving a region of vacuum located in $x > 0$ and a region of dielectric of dielectric constant ε located in $x < 0$. At the interface a solution for a mode bound to the surface and propagating along the interface in the z-direction is obtained. The highlights of the process of solution is only brief outlined in the following discussions [5, 11].

In the region $x > 0$ above, the interface the bound state solution is given by

$$\vec{E}_> (\vec{r}, t) = \left[\frac{ik}{\alpha_>} E_>^0, E_>^1, E_>^0 \right] e^{[-\alpha_> x + i(kz - \omega t)]}, \tag{6.20a}$$

where

$$\alpha_>^2 = k^2 - \frac{\omega^2}{c^2}. \tag{6.20b}$$

As in the treatment of the acoustic Rayleigh wave, above the interface the fields decay in intensity in the bulk with separation from the interface so that it is required that $\alpha_> > 0$. Below the surface, in the region $x < 0$ the bound state solution in the dielectric medium is given by

$$\vec{E}_< (\vec{r}, t) = \left[-\frac{ik}{\alpha_<} E_<^0, E_<^1, E_<^0 \right] e^{[\alpha_< x + i(kz - \omega t)]}, \tag{6.21a}$$

where

$$\alpha_<^2 = k^2 - \varepsilon \frac{\omega^2}{c^2}. \tag{6.21b}$$

For Eq. (6.21) to characterize a decaying wave below the surface it is again required that $\alpha_< > 0$.

The solutions above and below the interface must be match at the interface by the standard electromagnetic boundary conditions. Applying these to the electric field components: from the continuity of the tangential components of the electric field at the interface it follows that

$$E_>^0 = E_<^0 \quad \text{and} \quad E_>^1 = E_<^1, \tag{6.22a}$$

and from the continuity of the normal components of the displacement field at the interface

$$\frac{1}{\varepsilon} = -\frac{\alpha_>}{\alpha_<}. \tag{6.22b}$$

In addition, there are boundary conditions on the magnetic fields. These fields are obtained from Faraday's law applied to the electric field forms in Eqs. (6.20) and (6.21). Applying the magnetic field boundary conditions: from the continuity of the tangential components of the magnetic field at the interface it follows that

$$[\alpha_> + \alpha_<] E_>^1 = 0 \tag{6.23}$$

which requires that $E_>^1 = E_<^1 = 0$. Consequently, the surface wave described by Eqs. (6.20) and (6.21) for propagation in the z-direction has only an x- and z-component. In this regard, it represents an excitation with both longitudinal and transverse components.

The condition determining the dispersion relation of the modes bound to and traveling along the surface is given as the solutions of Eq. (6.22b). For the case in which $\varepsilon = $ a constant, Eq. (6.22b) yields the dispersion relation [5]

$$k = \sqrt{\frac{\varepsilon}{1 + \varepsilon}} \frac{\omega}{c}, \tag{6.24}$$

and considering the forms in Eqs. (6.20) and (6.21) the dispersion in Eq. (6.24) requires that $\varepsilon < -1$. In particular, the dispersion in Eq. (6.24) and the choice of the negative permittivity is consistent with Eq. (6.22b) and the fact that $\alpha_>, \alpha_< > 0$.

In the case in which $\varepsilon(\omega)$ is frequency dependent the resulting dispersion relation is a little more complicated. As an important example of such a dispersive permittivity, consider the free electron gas model of a metal having a frequency dependence of the form

$$\varepsilon(\omega) = \varepsilon_\infty \left[1 - \frac{\omega_p^2}{\omega^2} \right]. \tag{6.25}$$

Here ω_p is the plasma frequency and ε_∞ is the high frequency limit of the dielectric function. Applying Eq. (6.25) in Eq. (6.22b), it follows from some algebra that [5]

$$\omega^2(k) = \frac{1}{2\varepsilon_\infty}$$
$$\left\{ (1 + \varepsilon_\infty) c^2 k^2 + \varepsilon_\infty \omega_p^2 - \sqrt{\left[(1 + \varepsilon_\infty) c^2 k^2 + \varepsilon_\infty \omega_p^2 \right]^2 - 4\varepsilon_\infty^2 c^2 k^2 \omega_p^2} \right\}. \tag{6.26a}$$

The form in Eq. (6.26a) gives the dispersion relation of the surface electromagnetic wave solution which is known as the surface plasmon-polariton mode. It is found to be a dispersion relation which is a bounded function in k.

In the limit that $k \to 0$, Eq. (6.26a) gives the limiting form

$$\omega \to ck, \tag{6.26b}$$

and the plasmon-polariton dispersion relation is bounded by the light line in vacuum. In the limit that $k \to \pm\infty$, Eq. (6.26a) gives the limiting form

$$\omega \to \sqrt{\frac{\varepsilon_\infty}{1 + \varepsilon_\infty}} \omega_P, \tag{6.26c}$$

which is an upper bound on the surface wave dispersion relation. Taking these limits into account, Fig. 6.1 presents a schematic representation of the typical surface plasmon-polariton dispersion relation found at a vacuum-metal interface.

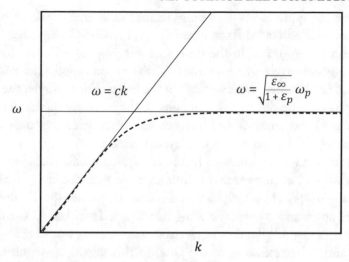

Figure 6.1: A schematic representation of a typical surface plasmon dispersion relation (dashed line) in regards to the limiting forms in Eqs. (6.26b) and (6.26c). The limiting lines $\omega = ck$ and $\omega = \sqrt{\frac{\varepsilon_\infty}{1+\varepsilon_\infty}}\omega_P$, which are upper bounds on the dispersion relation, are indicated on the figure by solid lines.

6.2.2   FINITE LAYERED MEDIA

Systems related to those of surfaces supporting electromagnetic waves have also been of recent interest. Specifically, these involve the electrodynamic excitations of layered mirror coatings and topics focused on the nature of plasmon-polariton excitations in layered guiding media [11, 30, 52–70]. As in the case of surfaces the interests in layered media encompass important applications in the technologies of a diverse range of devices. For the purposes of technology, the basic properties which make coatings of interest fall into two categories centered on the natures of the excitations within the coatings. These properties are the dispersive characteristics of the modes bound to the layers of the coatings and to the field intensity distributions within the layers. In the following these two basic features of coating excitations are discussed as they relate to technology.

A common element of many applications of layered coatings is the ability of the coating to concentrate the intensity of the electromagnetic fields within the volume of the coating layers and at the surfaces on which they are deposited. In particular, the ability to concentrate the field intensities is often found to be helpful in enhancing field dependent physical effects. In addition, other important applications are dependent on the nature of the dispersive propagation in the layers of the coating [11, 30, 52–70]. Here the different characteristics of bulk and surface wave propagation become important for the development of new metamaterials with exotic properties and for their applications or for the design of guiding structures.

Some examples of applications of coatings formed of layered media on mirrors have been introduced in schemes for enhanced Raman scattering [11, 53–56], the enhancement of second harmonic generation [57–63], and in the design of imaging [64–67] and focusing technologies [68, 69] offering subwavelength resolution. These various applications in general rely on the concentration of field intensities at surfaces in the layers. For example, the Raman and second harmonic effects operate directly as functions of the intensity of the fields present in the layering. On the other hand, imaging and focusing effects are related to the intensity as well as to the dynamical effects of the excitations at a layered surface.

In addition, important technologies involving layered media include the design of efficient waveguides [30] and in more recent technologies of new dispersive materials, known as hyperbolic metamaterials [7, 11], which have highly anisotropic transport properties [70, 71] allowing them to find applications in subwavelength imaging. These last applications focus on the ability of layered media to guide the flow of electromagnetic energy or to change the frequency-dependent propagation characteristics of the flow. In this regard, waveguides and hyperbolic media provide another set of systems in which the nature of the field distribution within the guiding media can be of considerable interest but for different reasons than solely for field enhancement of a physical effect [7, 11]. All these technological applications, however, are based in some part on the nature of the electromagnetic fields of the surface electromagnetic modes and how they are concentrated or dispersed throughout the layered medium at a variety of different ranges of the wavelength of light.

In this section, discussions are made of the field distributions in layered coatings on mirrors and within waveguide structures and hyperbolic media, with a focus on how they are generated, and how they are characterized and quantified [11]. In this regard, to facilitate the design of useful structures it is important to develop an understanding of the factors which determine the distribution of field intensities within the coating. Of interest are the factors that can concentrate the fields near the outer surfaces of the layered coating, within the center of the layered coating, or which will generate a uniform distribution throughout the layers. Such considerations are of interested in the study of enhanced Raman spectroscopy in which the spectroscopic enhancement arises from the intensity of the fields at the sites of the molecules being examined [11, 53–56]. These molecules may be located at a variety of sites within the layered system and their spectroscopic response is intimately related to the intensity of the frequency-dependent fields at their locations. Likewise, field enhancements can be of importance in the development of second harmonics generation [11, 57–63]. In these systems the intensity of the second harmonic fields is roughly quadratically dependent on the intensity of the fundamental fields applied to the nonlinear medium. Again, the nonlinear generating media can be located at a variety of different sites within the layered coating. In an important point, the total generated second harmonic fields from the sample are also related to the position at which the fields are generated in the sample and the phase addition of these fields [11]. Similarly, localizing effects are important in subwavelength focusing of near field microscopy and in the development of technologically useful metamaterial surfaces and media [11, 30, 64–71].

In light of the many different possible applications and the diverse nature of the systems considered, two very general models of a layered deposition are discussed here. These are chosen to be simple models which illustrate many of the properties found in more complex systems. Consequently, the models incorporate only the roughest general features of many of the systems mentioned earlier. Specifically, for a first model the considerations are for a deposition in a slab waveguide structure which also serves as a model for the anisotropic dispersive properties of a hyperbolic medium [11, 43, 70, 71]. Following these discussions a consideration is made of a layered deposition on a mirror [11]. Layered coatings on mirrors have found applications in second harmonic generation and surface enhanced Raman effects.

In a first set of considerations a presentation is given of a layered medium between a set of two perfectly conducting parallel plates. This is a model of a waveguide structure. In the case that many such elements are stacked on top of one another, they contribute to the formation of a crude model for a type of hyperbolic metamaterial whose technological properties are based on the spatial anisotropy of their modal dispersion relations [11]. For this system the modes transporting energy within the layered media in the plane of the plates are important. It is shown how these systems exhibit a variety of modes depending on the frequency of the radiation and that the modes exhibit a variety of field intensity distributions within the plane perpendicular to that of the perfectly conducting plates.

In a second set of considerations a treatment is presented of a layered medium coating deposited on a perfectly conducting planar mirror. In this system the surface wave modes transporting energy in the plane parallel to the mirror surface within the coating media are discussed. It is again found that a variety of modes exist depending on the frequency of the radiation and that these modes exhibit a variety of field intensity distributions within the plane perpendicular to that of the perfectly conducting mirror.

In the course of the treatment of these structures, a number of measures are discussed which can be used to characterize the field distributions of the modes in the direction perpendicular to the perfect conducting planes. These include plots of the field distributions, the mean and variance of the field distribution, and the entropy of the field distribution [11, 72, 73]. While all these measures are used in the following, a focus of the presentation is on the entropy of the intensity distribution of the modal fields. This is seen to give a most efficient characterization of the nature of the intensity distribution of the modes in space.

The entropy is a mathematical concept for the measurement of the broadness of a probability distribution [11, 72, 73], and in terms of the distribution of electromagnetic fields in space provides a useful characterization of the field localization within a layered medium. As such, the entropy measure is explained and used to characterize the distributions of the modes in the earlier described models. In this regard, a point in the following is a comparison of the various measures applied to the characterization of the localized nature of the fields within coating and layered structures.

Models

Waveguide and Metamaterial

The first system considered is a type of waveguide formed as a layered medium between two perfect conducting parallel plates, with the layering composed of slabs having interfaces in planes parallel to those of the perfect conducting plates. To develop a basic understanding of the problem, a simple model is treated in which the layering is composed of altering vacuum-dielectric slabs. (A schematic representation of the waveguide structure is given in Fig. 6.2a.) In addition, all the slabs in the layering are taken to be of thickness d and the dielectric slabs are set to have a homogeneous isotropic dielectric constant ε. For a specific illustration of the basic properties of such waveguides, the properties of a system of five dielectric slabs is presented.

A realization of the waveguide system of interest is presented in Fig. 6.2a. Here the x-axis of the space coordinates is in the vertical direction and the z-axis is in the horizontal direction. The perfect conducting parallel plates are located at the top and bottom of the schematic at $x = 0$ and $x = 11d$. Within the region between the prefect conducting plates are the vacuum-dielectric layers with the lower surfaces of the dielectric slabs located at $x = d$, $3d$, $5d$, $7d$, $9d$ and the tops of the dielectric slabs located at $x = 2d$, $4d$, $6d$, $8d$, $10d$. Everywhere else in the unshaded regions of the diagram between the plates is vacuum.

Layered Coating on a Perfect Conducting Mirror

The second system considered is a layered medium coated on a perfect conducting planar mirror. In this coating a finite periodic array of alternating layers composed of slabs having interfaces in planes parallel to those of the perfect conducting mirror is deposited on the perfect conducting surface. For a simple model results for a coating of vacuum-dielectric slabs are presented. This provides a basic model which qualitatively represents a variety of interesting properties of coated mirrors.

A schematic for the simple system of altering vacuum-dielectric slabs is presented in Fig. 6.2b. The slabs layered above the prefect conducting mirror base are of thickness d, and the dielectric slabs, represented as the shaded slabs, have a homogeneous isotropic dielectric constant ε. The regions of vacuum above the prefect conducting surface are unshaded. As in the waveguide discussions, for a specific illustration of the basic properties of coated mirrors, the properties of a system of five dielectric slabs are presented.

For the model in Fig. 6.2b the space coordinates are defined with the x-axis in the vertical direction and the z-axis is the horizontal direction. For these coordinates the perfect conducting mirror is located at $x = 0$. In addition, the dielectric slabs (shaded in the figure) have lower surfaces at $x = d$, $3d$, $5d$, $7d$, $9d$ and their top surfaces are at $x = 2d$, $4d$, $6d$, $8d$, $10d$. Outside the shade regions, everywhere else is vacuum.

Perfect Conductor

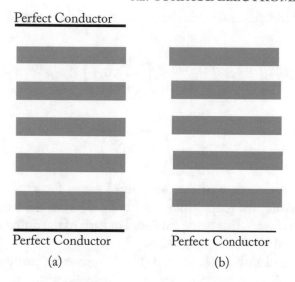

Perfect Conductor   Perfect Conductor

(a)   (b)

Figure 6.2: Schematic drawings of the two systems treated: (a) the waveguide composed of five dielectric layers (regions of grey) between two perfect conduction plates and (b) the coating composed of five dielectric layers (regions of grey) layered on a perfect conducting mirror. Both models are translationally invariant in the horizontal plane and the white regions are vacuum.

Surface Wave Modes

Within the layers of the layered media of both the waveguide in Fig. 6.2a and mirror coating in Fig. 6.2b, the electric field is a solution of Maxwell's equations. The general form of these fields in both systems in Fig. 6.2, for solutions bound in the layers and propagating along the layers in the z-direction, exhibits an exponential spatial dependence in the direction normal to the planar interfaces (i.e., the x-direction) and a planewave dependence in the z-direction. Consequently, the solutions within the vacuum layers are given by

$$\vec{E}(x,z,t) = \left[\left(\frac{ik}{\alpha}, 0, 1 \right) E^0_{v,z} e^{-\alpha(x-x_L)} + \left(-\frac{ik}{\alpha}, 0, 1 \right) E^1_{v,z} e^{\alpha(x-x_L)} \right] e^{i(kz-\omega t)} \quad (6.27a)$$

where

$$\alpha = \sqrt{k^2 - \left(\frac{\omega}{c} \right)^2} \quad (6.27b)$$

determines the exponential spatial dependence for $k^2 > \left(\frac{\omega}{c} \right)^2$. In Eq. (6.27), x_L is the x-coordinate of the lower edge of the particular vacuum slab being studied and k characterizes the propagation along the z-direction.

Similarly, considering the dielectric slabs the electric field in the dielectric layers are given by

$$\vec{E}(x, z, t) = \left[\left(\frac{ik}{\beta}, 0, 1 \right) E^0_{d,z} e^{-\beta(x-x_L)} + \left(-\frac{ik}{\beta}, 0, 1 \right) E^1_{d,z} e^{\beta(x-x_L)} \right] e^{i(kz-\omega t)}, \quad (6.28a)$$

where

$$\beta = \sqrt{k^2 - \varepsilon \left(\frac{\omega}{c} \right)^2} \qquad (6.28b)$$

determines the spatial dependence of the exponential forms for $k^2 > \varepsilon \left(\frac{\omega}{c} \right)^2$ in terms of the wavevector k characterizing the propagation. In Eq. (6.28) x_L is the x coordinate of the lower edge of the particular dielectric slab being studied. For each of the different slabs in Fig. 6.2 the x_L will assume a different value.

The forms presented in Eqs. (6.27) and (6.28) represent a particular set of solutions in which the electric field intensities propagate along the z-direction and are non-propagating along the x-direction. In the following, the distribution of field intensities of solution of the forms in Eqs. (6.27) and (6.28) are discussed. It should be noted that other solutions are possible, and the study of these additional modes proceeds much in the same way as the treatment of the bound modes with exponential forms perpendicular to the layer interface which are studied here.

The coefficients $\left\{ E^0_{v,z}, E^1_{v,z}, E^0_{d,z}, E^1_{d,z} \right\}$ in the solution presented in Eqs. (6.27) and (6.28) are determined for each of the slabs in Fig. 6.2 by an application of the electromagnetic boundary conditions. Between the vacuum and dielectric slabs, the electric field components parallel to the slab interfaces are continuous, the normal components of the electric displacement fields are continuous, and the electric field components parallel to the perfect conducting plates are zero at the surfaces of the perfect conducting plates. Similarly, considering the magnetic fields which are obtained by applying Faraday's law to the electric fields of Eqs. (6.27) and (6.28), the magnetic field components perpendicular to the dielectric-vacuum interfaces are continuous, and the magnetic field components perpendicular to the perfect conducting plates are zero at the perfect conducting plates.

The two systems studied here, which are represented in Fig. 6.2a and Fig. 6.2b, differ only in the boundary conditions at the top of the upper slab of the layering. In particular, for the case of the system in Fig. 6.2a the waveguide is bounded above by a perfect conducting plate whereas in Fig. 6.2b for the mirror coating there is no upper perfect conducting plate. In the waveguide geometry the electric field component tangent to the perfect conducting plate must vanish, but in the perfect mirror system the fields of the solutions which are bound to the coating must exponentially decaying in space with increasing upward separation from the coating layers. The fields of the coating, consequently, become zero at infinity.

For both systems in Fig. 6.2, once the fields are determined, the modulus squared of the electric field, $|E(x, z)|^2$, can be computed for each of the modal solutions obtained from the boundary conditions. This quantity is closely related, for example, to the transition rates commonly occurring in studies of surface enhanced Raman spectroscopy and in the generation of second harmonics of radiation in an appropriate nonlinear medium [1]. In the systems discussed here the slabs are considered to leading order to be linear dielectrics, and the response of a second harmonic generating nonlinearity or inelastic spectroscopy is made to the modal fields of the fundamental frequency within the layers. Corrections are not made for the losses to the modal solutions from the radiation generated by the weak nonlinearity in the system.

Characterization of the Electric Field Distribution in Space

A number of measures are available for the characterization of the spatial field intensity distribution, $|E(x, z)|^2$, within the layered medium. These treat the normalized intensity distribution as a probability distribution, providing a quantification of how the field intensity of the mode is distributed over the spatial extent of the layered media. In this way measures used to assess the fields distributed over space include the mean position, the variance of the distribution in position about the mean, and measures of the probability entropy of the field distribution in space [11, 72, 73].

As the layered system is translationally invariant in the z-direction, the field intensity only depends on x. Consequently, the intensity can be regarded as a type of probability distribution in x alone. The distribution of the fields in this probability density then determines the mean, variance, and various functions related to the entropy of the distribution in the x-coordinate. In particular, the differential entropy [74, 75], which is closely related to the entropy of the field distribution in space, will be used later to characterize the distribution of the fields within the layered media considered in Fig. 6.2. This will enable the assessment of where in the layered media are the fields most concentrated and where are they least concentrated. Such information is of interest in developing systems with functional properties related to field intensities concentrated within specific regions of the system.

The mean and the variance of the field intensity in the x-coordinate of the layers are defined in the standard way by

$$< x >= \frac{\int dx |E(x)|^2 x}{\int dx |E(x)|^2} \tag{6.29a}$$

and

$$var = \frac{\int dx |E(x)|^2 (x- < x >)^2}{\int dx |E(x)|^2}, \tag{6.29b}$$

respectively. In computing these averages the range of the integrals in x are limited to the regions over which the fields are defined, and the square root of the variance gives the standard deviation which is also commonly used as a characterization of the distribution. These probability measures provide the simplest quantification of the bulk distribution of the fields, i.e., are the fields concentrated at the center, top, or bottom of the slab. In addition, the standard deviation about the mean indicates how narrow or how broad is the field concentration in space.

Another means of characterizing the extent that the fields are distributed over the layered medium involves the entropy of the normalized field distributions positioned over the layers. The application of the entropy measure to the distributions provides an indicator of both the positioning and the degree of localization of the fields. In this regard, it provides a better classification of the nature of the fields than that supplied by the statistical moments of the field distribution.

First, the statistical nature of the entropy will be introduced and latter it will be applied to the fields of the layered media. As a general consideration, the entropy of a discrete set of probabilities $\{p_i\}$, normalized such that $\sum_i p_i = 1$, is defined by [11, 72–75]

$$S = -\sum_i p_i \ln(p_i). \tag{6.30}$$

It is a commonly used measure of the information content of a set of probabilities and has found many applications in the discussions of information and quantum information theory [72, 73]. It characterizes how much one learns on sampling a system described by the set of probabilities, i.e., the great the entropy of a complete set of probabilities the more information is extracted or gained from the results of a measurement on the system represented by the set of probabilities. In this regard, it can also be used to characterize how broadly a set of probabilities is distributed because more information is obtained from the measurements on systems which are broadly distributed over a variety of states than on systems limited in their range of possibilities.

In this last regard, if a sampling is made on a system consisting of one state which occurs with a probability of one, then Eq. (6.30) gives an entropy of zero. This value of entropy indicates that nothing new is learned in a measurement made on such a system. Since the distribution of probabilities is only over one sampling value, the system is always in that one state. On the other hand, for a system consisting of a number of different probability states, $\{p_i\}$, the entropy will be larger than zero. In this case, a measurement supplies information about the system, and the system is characterized by a distribution over various different possible outcomes. The greater the number of states the system is distributed over, the greater is the value of its entropy of the probability distribution and the more information is obtained from a measurement made on the system. Entropy is then a relative measure.

The entropy defined in Eq. (6.30) is for a discrete set of probabilities and our electric field is represented as a continuous distribution over space. The continuous nature of the distribution is not a problem as the definition of the entropy can be extended to continuous probability distributions in the form of the related differential entropy. The differential entropy measure is a straightforward generalization of the ideas developed in the study of entropy, but various differences between the two measures arise from the distinctions between discrete and continuous probability distributions. The extension of the ideas of entropy to treat continuum systems is now discussed.

A study of a continuous probability distribution $P(x)$ is transformed to that of the study of a set of discrete probabilities by discretizing the probability distribution using the trapezoidal rule. In this way the continuum probability variables $p(x)\,dx$ become a set of discrete probabilities of the form $\{P(x_i)\,\Delta x\}$. With this limit Eq. (6.30) is then rewritten in the form [74, 75]

$$S = -\sum_i P(x_i)\,\Delta x \ln\left(P(x_i)\,\Delta x\right), \tag{6.31a}$$

where $\{x_i\}$ is a set of consecutive discrete points and the nearest neighbor points are separated by Δx.

Letting $\Delta x \to 0$, Eq. (6.31a) then becomes [74, 75]

$$S = -\sum_i P(x_i)\ln\left(P(x_i)\right)\Delta x - \ln(\Delta x) \ \to\ -\int P(x)\ln\left(P(x)\right)dx - \ln(\Delta x). \tag{6.31b}$$

Notice that the term on the far right is an infinite constant which can be chosen to be the same for each discretized field distribution under study. This provides for a comparison of the properties of two different distributions defined over the same x-axis. Consequently, under these conditions of comparison, the differential entropy defined as [26, 27]

$$S_d = -\int P(x)\ln\left(P(x)\right)dx \tag{6.31c}$$

can be taken as a measure of the broadness of $P(x)$ in the variable x, distinguishing between two different probability distributions. In this scheme, broader distributions have higher positive values of S_d and narrower probability densities have more negative values of S_d.

As an example of the differential entropy measurement, consider the Gaussian distribution of the form [74, 75]

$$P(x) = \frac{1}{\sqrt{2\pi\sigma^2}}e^{-\frac{(x-\mu)^2}{2\sigma^2}}. \tag{6.32a}$$

Applying Eq. (6.32a) in the expression in Eq. (6.31c) for S_d, it follows that the differential entropy is given by

$$S_d = \ln\left(\sigma\sqrt{2\pi e}\right). \tag{6.32b}$$

From the result in Eq. (6.32b) it is seen that in the limit of a broad distribution in space $\sigma \to \infty$ and $S_d \to \infty$, and in the limit of a narrow distribution in space $\sigma \to 0$ and $S_d \to -\infty$. Consequently, as the width of the distribution increases in x, the differential entropy becomes increasingly more positive relative to the entropies of narrower distributions. Conversely, very narrow distributions have large negative values of differential entropy.

Another example of the differential entropy measurement is provided by the Poisson distribution. Consider a Poisson distribution of the general form

$$P(x) = \alpha e^{-\alpha x}. \tag{6.33a}$$

Upon substituting in Eq. (6.31c) it is found that this distribution has a differential entropy given by

$$S_d = 1 - \ln \alpha. \tag{6.33b}$$

In the limit of a narrow distribution in space $\alpha \to \infty$ and $S_d \to -\infty$, and in the limit of a broad distribution in space $\alpha \to 0$ and $S_d \to \infty$. Consequently, as the width of the distribution increases in x the differential entropy becomes increasingly more positive relative to the entropies of narrower distributions. For both distributions in Eqs. (6.32a) and (6.33a) the differential entropies of the distributions offer an effective quantification of the distribution of the probabilities over space.

Another example of a probability distribution which illustrates a useful feature of the entropy measure is now discussed. This is an example of a probability distribution which is highly localized about two different separated regions of space. For example, consider the stepwise constant distribution defined by [74, 75]

$$P(x) = \frac{1}{2h} \quad \text{for} \quad 0 \le x \le h \quad \text{and} \quad x_0 \le x \le x_0 + h \quad \text{but} \tag{6.34a}$$
$$P(x) = 0, \quad \text{otherwise}, \tag{6.34b}$$

where the form in Eq. (6.34) for $h \ll x_0$ consists of two different narrow localized regions separated from one another by an arbitrary distance, x_0. The distribution is over a few states which are separated from each other by a large separation, but it should not be considered as a broad distribution over many states.

First, consider studying the system using the standard variables of the mean and the standard deviation of the distribution. From Eq. (6.29a) the average position and from Eq. (6.29b) the standard deviation about the average position are, respectively,

$$< x > = \frac{1}{2}[h + x_0], \tag{6.35a}$$

and

$$Standard\ Deviation = \frac{1}{2}\sqrt{\frac{1}{3}h^2 + x_0^2}. \tag{6.35b}$$

Notice that for small h and large x_0 both the average position and the standard deviation become large, just as they would be for a broad distribution ranging over many possible states.

The localized nature of the distribution is missed in this particular statistical approach as the separation of the peaks, x_0, dominates the measures. It will now be shown that a better measure of such a distribution is given by the differential entropy.

The differential entropy of the distribution in Eq. (6.34) is obtained by substitution into Eq. (6.31c). This gives a differential entropy of the form

$$S_d = \ln(2h). \tag{6.36}$$

Now, unlike the characterization in Eq. (6.35), it is seen that the entropy is independent of the peak separation x_0 and only depends on the peak widths, h. For large values of h the entropy tends to large positive values and for small values of h the entropy tends to increasingly negative values of the entropy. Consequently, distributions consisting of separate narrow peaks are correctly characterized as representing localized field distributions. In this sense the differential entropy provides a much better characterization of the probability density than does the standard deviation.

Characterization of the Layered Media Systems Discussed Earlier

In the following, the theory just developed for the distribution of fields in space is illustrated by a treatment of the two layered media models in Fig. 6.2. The spatial properties of the modal solutions in these two models are studied for systems in which $\varepsilon = -2$ and $\varepsilon = -9$. These two different dielectric constants provide results that may be viewed as indicators of the qualitative behaviors in coatings with low and high contrasting dielectric layers. For the presentation the reader is reminded that both dielectric constants are negative in order to support bound modes at the vacuum-dielectric interfaces of the layers.

In the discussions that follow, calculations are presented for the dispersion relations of the modes propagating in the plane of the layer interfaces; the spatial dependence of the field intensities of the propagating modes; and the mean, variance, and entropy of the field distributions of the modes. The focus is on quantifying the spatial distribution of field intensity within the coatings of the various dispersive modes in the systems. First a discussion of the dispersive properties of the coating modes are given, and this is followed by a treatment of the spatial dependence of the field properties of these modes. Later some presentation is made for systems containing impurity layers.

In Fig. 6.3 results are presented for the dispersion relations of the modes propagating parallel to the planes of the layer interfaces. These dispersive curves are shown as plots of $\frac{\omega}{c}d$ vs. kd. The plots are made for both the waveguide model in Fig. 6.2a and for the coated mirror model in Fig. 6.2b, and it is generally found in both models that as kd is increased the number of modes in the systems increases. In the case of the coated mirror only modes bound to the coating are studied, and no discussion is made for bulk or leaky modes which occur at higher frequencies and propagate in the region of bulk above the mirror.

As an interesting point regarding the mirror and waveguide modes, the dispersive curves of the modes are seen to fall into two categories. Some of the modes have dispersion relations

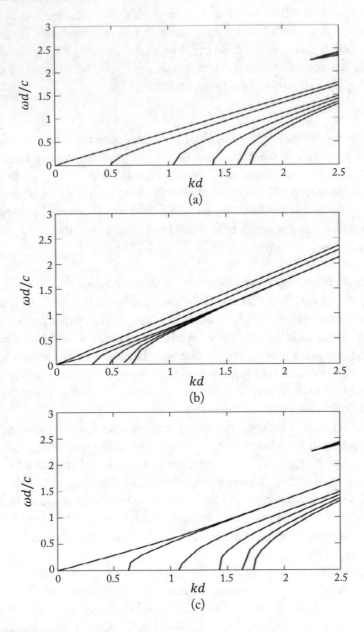

Figure 6.3: Plots of the dispersion relation $\frac{\omega}{c}d$ vs. kd for: (a) an $\varepsilon = -2$ and vacuum layering with the top layer open to a semi-infinite region of vacuum, (b) an $\varepsilon = -9$ and vacuum layering with the top layer open to a semi-infinite region of vacuum, and (c) an $\varepsilon = -2$ and vacuum layering between two perfect conducting plates.

which begin at $\frac{\omega}{c}d = 0$ and increase in frequency with increasing wave numbers. Other modes have dispersion relations which begin at non-zero frequencies and again increase in frequency with increasing wave numbers. These two different types of dispersive modes occur first at low frequencies in systems with lower dielectric contrast.

The dispersive curves in Fig. 6.3a for the mirror coating are found to be similar to those in Fig. 6.3c for the waveguide. In both plots $\varepsilon = -2$, so that the only difference in the two systems is that one has an additional upper perfect conducting plate. In this regard, it is seen later that the wavefunctions in both of these models are fairly well confined to the region of the layered coating. The results in Fig. 6.3b, on the other hand, are for the mirror coating evaluated for $\varepsilon = -9$. In this system dispersive modes which start at non-zero frequencies are found to occur at higher wave numbers than those of the plots for $\varepsilon = -2$. The significant difference between Figs. 6.3b, 6.3a, and 6.3c is the difference in dielectric contrast in the coating.

For technological applications it is sometimes necessary to develop specific modal field intensities at particular frequencies and locations within the layers. This can be done by choosing the geometry and dielectric constants of the layers correctly and developing an understanding of the frequency and field distributions of the modal solutions within the layers. As an illustration of this point, a consideration is now given of the field distributions within the layered media in Fig. 6.2 and their realizations in the systems with dispersions given in Fig. 6.3.

In Fig. 6.4 results for the wave functions are presented for some of the modes in Fig. 6.3 considering the particular case in which $kd = 1.5$. The figures show the electric field intensity measured by $|E(x,z)|^2$ plotted as a function of x measured in units of d. (In the plots it should be remembered that the system is translationally invariant in z so that $|E(x,z)|^2 = |E(x)|^2$.) Results are shown for both the waveguide and layered mirror models, with results given for the mirror coating model in the cases that $\varepsilon = -2$ and $\varepsilon = -9$ in Fig. 6.4a and 6.4b, respectively, and for the waveguide model with $\varepsilon = -2$ in Fig. 6.4c.

Considering the intensity plots in Fig. 6.4, it is generally found that the spatial field intensity profiles of the modes fall into two types. These include modes with intensities concentrated at the top and/or bottom of the layered coating and modes with intensities uniformly distributed throughout the layering. For example, in Fig. 6.4a for the mirror coating geometry in Figs. 6.2b and 6.4c for the waveguide geometry in Fig. 6.2a there are modes localized at both the top and the bottom of the layering. For the mirror coating results presented in Fig. 6.4b, however, localized modes are found only at the top of the layering, away from the mirror surface. The location of the highly localized modes in these systems are greatly dependent on the value of the dielectric constants.

In addition to the highly localized modes just discussed, in all of the plots in Fig. 6.4 are also a series of modes with intensities spread throughout the layers. These modes occur at different frequencies than the highly localized modes and represent different frequency responses of the layers. The details of the modes presented in Fig. 6.4 and their occurrence in frequency

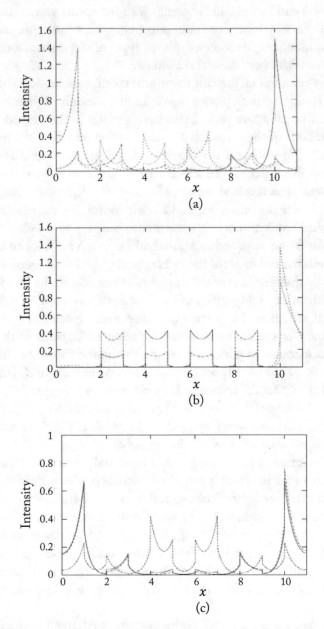

Figure 6.4: Plots of the electric field intensity vs. x in units of d for: (a) an $\varepsilon = -2$ and vacuum layering with the top layer open to a semi-infinite region of vacuum; (b) an $\varepsilon = -9$ and vacuum layering with the top layer open to a semi-infinite region of vacuum; and (c) an $\varepsilon = -2$ and vacuum layering between two perfect conducting plates.

are now discussed. In these discussions, the nature of the field distributions is correlated with their frequency occurrence.

In Fig. 6.4a the $kd = 1.5$ results are for a mirror coating with $\varepsilon = -2$. The modes of the coating are found to occur at a variety of discrete frequencies $\frac{\omega d}{c} = 0.3822,\ 0.69135,\ 0.93735,$ and 1.06069 and encompass both extended and highly localized space modes defined over the layers. Two modes with highly localized field intensities in space are present, one localized mode at the top of the coating and the other localized near the surface of the mirror. The mode localized at the top of the layers has frequency $\frac{\omega d}{c} = 1.06069$, and the mode localized at the bottom of the layers near the mirror surface has $\frac{\omega d}{c} = 0.93735$. These are the highest frequency modes of the system, and their excitation in the layers would generate concentrated fields with the modal frequency at either the top or bottom of the layered array. The wave functions of the remaining frequency modes of the system differ from these localized modes in having intensity distributions which are spread out over all of the layers. These remaining nonlocalized modes occur at much lower frequencies from those of the highly localized modes.

For the plot in Fig. 6.4b the system is also a mirror coating but now for the $kd = 1.5$ case in which $\varepsilon = -9$. The coating modes for this higher dielectric contrast occur at $\frac{\omega d}{c} = 1.16367,\ 1.17089,\ 1.304769$, and 1.4142135. From the plot it is found that two modes with highly localized field distributions exist, but now both localized modes are concentrated at the top of the coating and away from the surface of the mirror. One of the modes localized at the top of the coating is at the frequency $\frac{\omega d}{c} = 1.4142135$ and the other mode localized at the top of the coating is at the frequency $\frac{\omega d}{c} = 1.304769$. Both localized modes, however, have different intensity peak profiles, and this accounts for their frequency differences. As with the results in Fig. 6.4a, the remaining modes of the system have intensity distributions which are spread out over all the layers and are found at lower frequencies than the highly localized modes in the system.

In the plot in Fig. 6.4c the $kd = 1.5$ modes are for a layered slab waveguide with $\varepsilon = -2$ and perfect conduction top and bottom plates. With these conditions the waveguide modes are found to occur at the frequencies $\frac{\omega d}{c} = 0.3555,\ 0.6765,\ 0.9300$, and 0.9438. Due to the symmetry of the system along the vertical axis in Fig. 6.2a, the guided modes at these frequencies exhibit spatial intensity distributions which are symmetric about the center plane of the layered medium. Consequently, from their plots it is seen that a set of modes exist with high field concentrations localized at both the top and the bottom of the layering. One highly concentrated localized mode at the top and bottom of the layers is at frequency $\frac{\omega d}{c} = 0.9300$ and the other mode highly localized at the top and bottom of the layers is at frequency $\frac{\omega d}{c} = 0.9438$. These are the two highest frequency modes of the system. As in the mirror coating system, the lower frequency modes are spread throughout the layered media.

An interesting feature of the results presented for the waveguide and coated mirror modes at fixed wavevector is that the localized modes appear to be the high frequency modes of the system. The remaining lower frequency modes tend to have field intensities distributed evenly throughout the layers of the coating. This feature related to the mode frequencies arises from the contributions of the spatial derivatives entering the modal eigenvalue problem. Specifically, modes with a high space localization have high derivative contributions to their frequency eigenvalues.

Statistical Characterization of the Modes

Plotting the wavefunctions of all the modes is seen to be a tedious way of characterizing their spatial intensity distributions. A simpler characterization is provided from the statistical parameters obtained by treating the intensity distributions as probability distributions. The statistical parameters used for this characterization include the mean value $< x >$ of the vertical position in the layers, the standard deviation about this position (defined as the square root of the variance defined in Eq. (6.29b)), and the statistical entropy defined in Eq. (6.31c). In the following, this approach is applied to the results for the mirror coating model in Fig. 6.2b.

In Fig. 6.5 statistical results are presented considering the system in Fig. 6.2b for the case in which $\varepsilon = -2$. Results are shown for the mean value $< x >$, the standard deviation about this position, and the statistical differential entropy defined in Eq. (6.31c), all plotted as functions of kd. (Note: The wavefunctions for the system at the particular value $kd = 1.5$ in this study are presented in Fig. 6.4a.) The plots offer a simple quantitative measure of the spatial properties of the field intensities evaluated over the large range of modal states.

In Fig. 6.5a results are presented for $< x >$ as a function of kd. The crosses (+) at the top of the plot are for the high frequency localized modes occurring at the upper surface of the mirror coating. On the other hand, the (x) are for the high frequency localized modes which for $kd > 1$ are localized about the lower layers closest to the perfect conducting mirror. Notice that for $kd < 1$ both of the average positions of these localized modes approach the surface of the layering which is farthest from that of the mirror. A series of modes at $kd > 1$ occur near $< x > \approx 6$. These are two broad modes which are centered about the middle of the layering.

The standard deviations of the modes in Fig. 6.5a are presented in Fig. 6.5b. The localized modes plotted in Fig. 6.5a with the (+) and (x) and the results for their standard deviation in Fig. 6.5b are correlated in the (+) and (x) notation. Note that the (+) and (x) modes become more tightly localized in space as kd increases. For $kd \leq 0.5$ the standard deviation of the (+) and (x) modes become of order of the standard deviations of the broad modes which occur in the upper-most two rows of points in the region $kd > 1.5$. Also notice that in the region $kd > 1.5$ the number of broad modes in the system increases.

In Fig. 6.5c the differential entropy of the modes in Fig. 6.5a is presented. The localized modes plotted in Fig. 6.5a with the (+) and (x) and the results for their entropy in Fig. 6.5b are correlated in the (+) and (x) notation. Notice that the (+) and (x) modes become more tightly localized in space as kd increases. For $kd \leq 0.5$, the entropy of the (+) and (x) modes becomes of order of the entropy of the broad modes which occur in the upper-most two rows of points in the region $kd > 1.5$. As an additional point, in the region $kd > 1.5$ the number of broad modes in the system increases.

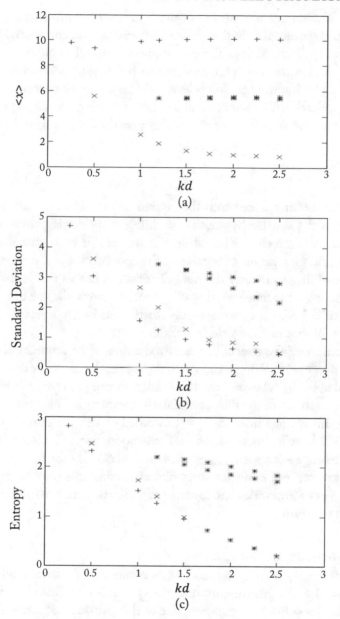

Figure 6.5: Plots for an $\varepsilon = -2$ and vacuum layering with the top layer open to a semi-infinite region of vacuum presenting: (a) the average value of x for the field distribution of the modes as a function of kd; (b) the standard deviation of the value of x about its average for the field distribution of the modes as a function of kd; and (c) the entropy for the field distribution of the modes as a function of kd. The x is measured in units of d, and points are shown for $0.25 \le kd \le 2.5$.

The results of the statistical measures of mode localization presented in this section show that the nature of the spatial distribution of field intensities are represented by basic statistical measures of probability distributions. These measures are useful to quickly compare the development of the field localization in frequency and with changing wavevector of the bound and guided modes. They provide information relevant to the design of coatings and guiding media for technological applications. As a further illustration a coating containing an impurity layer is now discussed, focusing on the effects of the impurity on the field distribution within the layers.

Layers with Impurities

The Impurity Model

An impurity slab can be introduced into the layered coating in Fig. 6.2b by replacing one of the dielectric slabs of the coating by a slab of a different dielectric constant from that of the original slab. This allows for a change in the electric field profile in the coating along the vertical axis in Fig. 6.2b, providing a means of regulating the concentration of field intensity within the mirror coating. The ability to modify the field distribution in this way is helpful in the design of coatings for a variety of technological applications. In particular, factors which are used to tune the intensity of fields as they are spatially distributed in the layers facilitate applications dependent on these distributions of field intensities.

As an illustration of the effects of the introduction of an impurity slab into the mirror coating, a study of the case involving a change in the slab at the center of the coating in Fig. 6.2b is given. In particular, in the following the third slab up from the mirror in Fig. 6.2b is replaced by an impurity slab so that ε of the dielectric slabs becomes $\varepsilon_{impurity}$ for the impurity slab.

Considering this change from the system studied earlier, the pure system mirror coating with values of $\varepsilon = -2$ have been calculated for the introduction of an impurity with a dielectric constant within the range $-2.6 \leq \varepsilon_{impurity} \leq -1.4$. This allows for the treatment of systems with impurities having greater and lesser impurity dielectric constants than that of the host system and provides examples of impurities with a range of dielectric constants about those of the other slabs in the layered medium.

Changes in the Field Statistics Due to an Impurity

The results for the impurity coating changed in this manner are presented in Fig. 6.6 for the case of modes with $kd = 1.5$. At this wave number the system is seen from Fig. 6.3a to exhibit four distinguishable modal solutions. In addition, due to the translational symmetry along the horizontal axis in Fig. 6.2b, the properties of these modes are only renormalized by the introduction of the dielectric impurity. In particular, they maintain their characterization as modes of wave number $kd = 1.5$, and the basic forms of the modal wave functions and dispersions are similar to those of the unperturbed system.

In Fig. 6.6a the changes of the frequency, $\frac{\omega d}{c}$, of the $kd = 1.5$ modes are presented as functions of the small variations of the impurity dielectric about the $\varepsilon = -2$ value of the pure

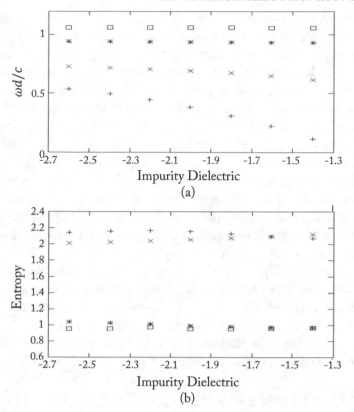

Figure 6.6: Plot of: (a) $\frac{\omega}{c}d$ vs. $\varepsilon_{impurity}$ and (b) the differential Entropy vs. $\varepsilon_{impurity}$ for the impurity slab system discussed in page 180. The point types used in the two plots correlate between the sets of data in (a) and (b).

layered coating. The two higher frequency modes of the system are seen to exhibit little variation in their frequencies with changes in the impurity dielectric. The two lower frequency modes of the system exhibit an increase in frequency with increasingly more negative $-2.0 > \varepsilon_{impurity}$. On the other hand, the two lower frequency modes of the system exhibit a decrease in frequency with less negative $-2.0 < \varepsilon_{impurity}$.

The feature of interest here is the effect of the impurity on the distribution of the fields along the vertical axis of the slabs. This is measured by the differential entropy of the intensity distributions of the fields in space.

In Fig. 6.6b the differential entropy of the four $kd = 1.5$ modes is presented as a function of small variations of the impurity dielectric about the $\varepsilon = -2$ value of the pure layered coating. Changes in the entropy with this variation are primarily seen in the broad modes, and the symmetry of these modes is also a consideration in setting the magnitude of the entropy change.

Since the impurity slab is located at the center of the layering, it is not surprising that the major effects of the impurity are on the broad modes which are distributed throughout the layering. These have significant intensities near the center of the coating. The other modes of the system have been shown to be localized either at the top or bottom of the coating, and, consequently, are affected less by the impurity which is positioned at the center of the coating.

In Fig. 6.6b the two lower frequency modes of the system are the broad modes with high entropies. They are found to exhibit opposite variations in their frequencies with changes in the impurity dielectric constant. For decreasingly negative $-2.0 < \varepsilon_{impurity}$, one mode increases in entropy while the other decreases. This difference arises from the detailed spatial dependence of the electric fields along the vertical axis and how the electric fields evolve from the fields of the parity states in the waveguide problem. The symmetry of the waveguide system is broken by removing its top perfect conducting plate, but the mirror coating states remember this now broken symmetry. For increasingly negative $-2.0 > \varepsilon_{impurity}$, again one mode increases in entropy while the other exhibits a small decrease in entropy.

In comparison, the two higher frequency modes of the system exhibit a decrease in entropy with increasingly more negative $-2.0 > \varepsilon_{impurity}$. These higher frequency modes of the system also exhibit an increase in entropy with less negative $-2.0 < \varepsilon_{impurity}$. Similar factors to those affecting the low frequency modes are at play with the high frequency modes.

In both sets of modes an increasing of the modal entropy indicates a broadening of the electric field intensity as distributed in space.

6.3   APPLICATIONS OF ELECTROMAGNETIC AND ACOUSTIC SURFACE WAVES

The discussions in this chapter have focused on the nature of electromagnetic and acoustic surface waves for both surfaces and surface layerings or coatings. The interest in these arises from a variety of applications in technology. In the following, some of these will be noted along with references of their more detailed, advanced, treatments in the literature. First electromagnetic surface waves are discussed. This is often classified as part of the general field of plasmonics. Following these considerations, some remarks on surface acoustic wave applications are made.

As has been mentioned earlier, two important applications of electromagnetic surface waves are surface enhanced Raman spectroscopy [53–56] and the generation of higher harmonics of radiation [57–63] at surfaces. Both of these applications rely on excited surface modes which create enhanced frequency-dependent fields at the surface sites of the atoms and molecules or at the optical nonlinear medium. The surface excitations forming the basis of these spectroscopies, in turn, are generated on the interface by bulk waves incident on the interface or by means of optical spasers situated on the interface. Eventually, after their processing on the interface, the surface waves are then reconverted to outgoing bulk waves propagating from the surface.

In these processes, bulk waves can be made to excite surface waves by either introducing disorder on the interfaces supporting the surface waves or by introducing optical prisms in their proximity [76]. In this regard, rough surface scattering mixes the bulk and surface modes while placing a prism close to the surface supporting surface waves is also a means of coupling bulk waves to the surface modes on the interface. Both processes are common mechanisms for scattering the two types of modes into one another and may also act to scatter surface waves into radiating bulk waves.

Another way of exciting surface wave modes is by spaser [77]. Electromagnetic spasers on the other hand are a type of nanoparticle based lasing mechanism which outputs coherent surface plasmons on plasmonic surfaces. The mechanism upon which spasers function is, in fact, another interesting application of electromagnetic surface waves and the principles of laser operations.

Other recent applications of plasmonics are in the design of various waveguides and circuits for the channeling of the flow of plasmonic energy [78]. This is a topic of optoelectronics and has a focus of replacing electronic technology by optical replacements. Various types of plasmonic devices have been designed based on these technologies.

Plasmonics has also found applications in medicine and biology. One of the early applications of plasmons to explain optical phenomena was in some of the optical properties of colored glasses [11]. Specifically, it was noted that the introduction of small gold particles to certain glasses caused them to exhibit a red color. The red color comes from the excitation of plasmons on the gold particles. A similar phenomenon has been suggested for medical applications. In particular, as an example, if certain types of nanoparticles are selectively absorbed in different types of body tissues, the excitations of plasmons in these particles can be used to diagnose the properties of the absorbing tissues [79].

As mentioned earlier, hyperbolic metamaterials formed from metal-insulator layers can also be used to create systems with hyperbolic dispersion relations [43, 70, 71]. In wavevector space the constant frequency surfaces of light in a homogeneous medium with dielectric constant $\varepsilon > 0$ are spheres. Consequently, in the far field optics of such a material the resolution at this frequency is limited by these propagating wavevector components. In a hyperbolic metamaterial, however, in wavevector space the constant frequency surfaces of light are hyperbolas. This is a far less restricting medium for the propagation of light and the far field optics can provide a much greater resolution. (Notice that while in a homogeneous medium the propagating wavevectors are bounded, the propagating wavevectors in hyperbolic metamaterial are unbounded.) This offers a great potential for far field subwavelength optical resolution in hyperbolic metamaterials. In this regard, another realization of a hyperbolic metamaterial is that of an array of parallel axis wires embedded in an $\varepsilon > 0$ dielectric. This system also exhibits a type of hyperbolic dispersion relation for the propagation of light.

Other plasmonic device schemes employ surface waves for increased optical resolution related to near field optical phenomena [64–69]. These involve near field probes which scatter surface waves into bulk waves where they are collected to make a microscopy. In a related idea, the sensitivity of the properties of surface electromagnetic waves have also been proposed as the basis of various sensing devices [80]. These have found applications in chemical and biological systems.

Plasmonic applications have also aided in the design of devices for solar energy harvesting [81].

Similar ideas to those for applications of surface electromagnetic waves occur in the discussions of surface acoustic wave (SAW) devices which have been used as components in various electronic technologies [46]. These devices function on the basis of transforming electronic and acoustic signals between one another. They have formed the mechanism of applications such as filters, transformers, transducers, oscillators, sensors, etc.

References

[1] Marder, M. P., 2000, *Condensed Matter Physics*, John Wiley & Sons, Inc., New York. DOI: 10.1002/9780470949955. 1, 2, 4, 7, 9, 10, 11, 12, 13, 15, 16, 18, 22, 23, 55, 56, 57, 58, 59, 61, 62, 63, 70, 72, 73, 74, 75, 76, 77, 80, 81, 82, 85, 86, 88, 90, 97, 119, 169

[2] Kittle, C., 1996, *Introduction to Solid State Physics*, John Wiley & Sons, Inc., New York. 1, 2, 4, 7, 9, 10, 11, 12, 13, 15, 16, 18, 22, 23, 55, 57, 58, 59, 61, 62, 63, 70, 72, 73, 74, 75, 77, 90, 97, 119

[3] Berman, O. L., Lozovik, Y. E., Eiderman, S. L., and Coalson, R. D., 2006, Superconducting photonic crystals: Numerical calculations of the band structure, *Physical Review B74*, 092505. DOI: 10.1103/physrevb.74.092505. 56, 57, 58

[4] Solymar, L. O. and Walsh, D., 2010, *Electrical Properties of Materials*, 8th ed., Oxford University Press, Oxford, Chapter 15. DOI: 10.1093/oso/9780198829942.001.0001. 1, 2, 4, 6, 7, 8, 9, 10, 11, 61, 62, 63, 70, 72, 73, 74, 75, 77, 97

[5] Wallis, R. F., 1986, in *Electromagnetic Surface Excitations*, Springer-Verlag, Berlin, Chapter 1. DOI: 10.1007/978-3-642-82715-0. 153, 154, 155, 160, 161, 162

[6] Calcote, L. R., 1968, *Introduction to Continuum Mechanics*, D. Van Nostrand Company, Inc., Princeton. 1, 9, 10, 11, 12, 14, 15, 16, 17, 18, 19, 20, 21, 22, 23, 24, 25, 26, 37

[7] Deymier, P. A., 2013, *Acoustic Metamaterials and Phononic Crystals*, Springer, Berlin. DOI: 10.1007/978-3-642-31232-8. 1, 2, 3, 7, 10, 11, 12, 13, 14, 15, 16, 17, 18, 19, 20, 21, 22, 23, 24, 25, 26, 37, 97, 99, 100, 113, 114, 118, 121, 123, 127, 133, 139, 142, 150, 164

[8] Su, Y.-C., 2015, Design of elastic metamaterials, *Open Access Dissertations*, 710. 118, 121, 123, 127

[9] Pai, P. F. and Huang, G., 2015, *Theory and Design of Acoustic Metamaterials*, SPIE Press, Bellingham. DOI: 10.1117/3.2199731. 114

[10] Zhang, S., 2010, Acoustic metamaterial design and applications, Dissertation for Doctor of Philosophy in Mechanical Engineering, University of Illinois at Urbana-Champaign. 118

[11] McGurn, A. R., 2018, *Nanophotonics*, Springer, Cham. DOI: 10.1007/978-3-319-77072-7. 1, 2, 3, 4, 5, 6, 7, 8, 10, 61, 62, 69, 97, 98, 99, 101, 102, 111, 112, 113, 114, 115, 116,

118, 133, 134, 135, 136, 137, 139, 140, 141, 142, 143, 144, 146, 147, 149, 150, 154, 155, 160, 161, 163, 164, 165, 169, 170, 183

[12] Huang, H. H., Sun, C. T., and Huang, G. L., 2009, On the negative effective mass density in acoustic metamaterials, *International Journal of Engineering Science 47*, pp. 610–617. DOI: 10.1016/j.ijengsci.2008.12.007. 118, 119, 121, 134

[13] Huang, H. H. and Sun, C. T., 2011, Locally resonant acoustic metamaterials with 2D anisotropic effective mass density, *Philosophical Magazine*, 91(6):981–996. DOI: 10.1080/14786435.2010.536174. 119, 121, 122, 123, 134

[14] Li, J. and Chan, C. T., 2004, Double-negative acoustic metamaterial, *Physical Review E 70*, 055602–1 to 055602–4. DOI: 10.1103/physreve.70.055602. 127, 128, 129, 130, 131, 132, 139, 140

[15] Ding, Y., Liu, Z., Qiu, C., and Shi, J., 2007, Metamaterial with simultaneously negative bulk modulus and mass density, *Physical Review Letters 99*, 093904–1 to 093904-4. DOI: 10.1103/physrevlett.99.093904. 118, 127, 128, 129, 130, 131, 132, 140

[16] Movchan, A. B. and Guenneau, S., 2004, Split-ring resonators and localized modes, *Physical Review B 70*, 125116–1 to 125116-5. DOI: 10.1103/physrevb.70.125116. 115, 134

[17] Huang, H. H. and Sun, C. T., 2012, Anomalous wave propagation in a one-dimensional acoustic metamaterials having simultaneously negative mass density and Young's modulus, *Journal of the Acoustical Society of America 132*, pp. 2887–2895. DOI: 10.1121/1.4744977. 2, 3, 10, 113, 114, 118, 126, 127, 128, 129, 130, 131, 132, 133

[18] Shadrivov, H. V., Reznik, A. N., and Kivshar, Y. S., 2007, Magnetoinductive waves in arrays of split-ring resonators, *Physical B394*, pp. 180–183. DOI: 10.1016/j.physb.2006.12.038. 3, 137

[19] Giri, P., Choudhary, K., Gupta, A. S., Bandyopadhyay, A. K., and McGurn, A. R., 2011, Klein–Gordon equation approach to nonlinear split-ring resonator based metamaterials: One-dimensional systems, *Physical Review B 84*, 155429–1 to 155429-10. DOI: 10.1103/physrevb.84.155429. 1, 2, 3, 118, 137, 138

[20] Khelif, A., Aoubiza, B., Mohammadi, S., Adibi, A., and Laude, V., 2016, Complete band gaps in two-dimensional phononic crystal slabs, *Physical Review E74*, 046610–1 to 046610-5. DOI: 10.1103/physreve.74.046610. 1, 6, 7, 61, 62, 97, 99, 100, 101

[21] Joannopoulos, J. D., Johnson, S. G., Winn, J. N., and Meade, R. D., 2008, *Photonic Crystals: Molding the Flow of Light*, Princeton University Press, Hoboken, NJ. 1, 10, 97, 98

[22] McGurn, A. R., 2015, *Nonlinear Optics of Photonic Crystals and Meta-Materials*, IOP Concise Physics, Morgan & Claypool Publishers, San Rafael, CA. DOI: 10.1088/978-1-6817-4107-9. 1, 6, 7, 8, 10, 61, 62, 97, 113, 114, 118, 154

[23] Laude, V., 2015, *Phononic Crystals for Sonic, Acoustic, and Elastic Waves*, De Gruyhter, Inc. ProQuest Ebook Central. http://ebookcentral.proquest.com/lib/wmichlib-ebooks/detail.action?docID=4001494 and http://ebookcentral.proquest.com/lib/wmichlib-ebooks/detail.action?docID=4001494 97, 99

[24] McGurn, A. R. and Maradudin, A. A., 1993, Photonic band structures of two- and three-dimensional periodic metal or semiconductor arrays, *Physical Review B48*, pp. 17576–17579. DOI: 10.1103/physrevb.48.17576. 1, 6, 7, 61, 98, 100

[25] Shelby, R. A., Smith, D. R., and Schultz, S., 2001, Experimental verification of a negative index of refraction, *Science 292 (55140)*, pp. 77–79. DOI: 10.1126/science.1058847. 2, 3, 4, 5, 6, 113, 114, 115, 134, 139, 149

[26] Pendry, J. B., Negative refraction makes a perfect lens, *Physical Review Letters 85*, pp. 3966–3970. DOI: 10.1103/physrevlett.85.3966. 2, 115, 146, 147, 148, 149, 171

[27] Vesclago, V. G., 1967, The electrodynamics of substances with simultaneously negative values of ε and μ, *Soviet Physics Uspekhi*, 10(4):509–514. DOI: 10.1070/pu1968v010n04abeh003699. 2, 3, 4, 5, 6, 113, 114, 134, 139, 140, 171

[28] Royer, D. and Dieulesaint, E., 2000, *Elastic Waves in Solids I: Free and Guided Propagation*, Springer, Berlin. 2, 9, 10, 11, 12, 14, 15, 16, 17, 18, 19, 20, 21, 22, 23, 24, 25, 26, 35, 37, 40

[29] Faran, J. J., 1951, Sound scattering by solid cylinders and spheres, *Journal of Acoustic Society of America 23*, pp. 405–418. DOI: 10.1121/1.1906780. 2, 35, 36, 37, 38, 39, 40, 42, 43, 44, 45, 46, 47

[30] Jackson, J. D., 1975, *Classical Electrodynamics*, 2nd ed., John Wiley & Sons, Inc., New York. 3, 9, 10, 26, 27, 28, 29, 30, 31, 32, 33, 35, 47, 49, 50, 51, 53, 54, 55, 56, 102, 103, 104, 105, 106, 107, 108, 109, 110, 111, 113, 116, 160, 163, 164, 165

[31] Van de Hulst, H. C., 1981, *Light Scattering by Small Particles*, Dover Publications, Inc., New York, Chapter 15. DOI: 10.1063/1.3060205. 9, 10, 35, 47

[32] Arfken, G. B. and Weber, H. J., 2005, *Mathematical Methods for Physicists*, 6th ed., Elsevier Academic Press, Amsterdam. DOI: 10.1016/C2009-0-30629-7. 40, 41, 42, 49, 50, 51, 53, 54, 56, 62, 63, 64, 69, 70, 72, 73, 76, 77, 78, 79, 80, 81, 82, 85, 86, 88, 90, 91, 93, 97

[33] Tinkham, M., 2003, *Group Theory and Quantum Mechanics*, Dover Publications, Mineola. 62, 63, 64, 66, 70, 71, 72, 73, 75, 76, 77, 78, 79, 80, 81, 83, 85, 86, 89

[34] Herget, W. and Dane, M., 2003, Group theoretical investigations of photonic band structures, *Physica Status Solidi (a)*, 197(3):620–634. DOI: 10.1002/pssa.200303110. 79, 80, 81, 82, 85, 86, 88, 89, 90, 91, 93, 95, 96, 97

[35] Maradudin, A. A. and McGurn, A. R., 1993, Photonic band structure of a truncated, two-dimensional, periodic dielectric medium, *Journal of the Optical Society of America B10*, pp. 302–313. DOI: 10.1364/josab.10.000307. 98, 99

[36] McCall, S. L., Platzman, P. M., Dalichaouch, R., Smith, D., and Schultz, S., 1991, Microwave propagation in two-dimensional dielectric lattices, *Physical Review Letters 67*, pp. 2017–2020. DOI: 10.1103/physrevlett.67.2017. 98

[37] Poli, F., Cucinotta, A., and Selleroi, S., 2007, *Photonic Crystal Fibers: Properties and Applications*, Springer, Berlin. DOI: 10.1007/978-1-4020-6326-8. 101, 102

[38] Paschotta, R., 2010, *Field Guide to Optical Fiber Technology*, SPIE, Binghamtom, NY. DOI: 10.1117/3.853722.

[39] Shamonina, E., Kalinin, V. A., Ringhofer, K. H., and Solymar, L., 2002, Magnetoinductive waves in one, two, and three dimensions, *Journal of Applied Physics 92*, pp. 6252–6261. DOI: 10.1063/1.1510945. 113, 114, 115, 116, 117, 118

[40] Chang, H.-M. and Liao, C., 2011, A parallel derivation to the Maxwell-Garnett formula for the magnetic permeability of mixed materials, *World Journal of Condensed Matter Physics 1*, pp. 55–58. DOI: 10.4236/wjcmp.2011.12009. 115, 123

[41] Huang, H. H. and Sun, C. T., 2011, Theoretical investigation of the behavior of an acoustic metamaterial with extreme Young's modulus, *Journal of Mechanics and Physics of Solids 59*, pp. 2070–2081. DOI: 10.1016/j.jmps.2011.07.002. 115, 116, 117, 118, 123, 125, 126, 127, 128, 129, 131

[42] Markel, V. A., 2016, Introduction to the Maxwell Garnett approximation: Tutorial, *Journal of the Optical Society of America A33*, pp. 1244–1257. DOI: 10.1364/josaa.33.001244. 115, 118, 133, 136

[43] Caloz, C. and Itoh, A., 2005, *Electromagnetic Metamaterials: Transmission Line Theory and Microwave Applications*, Wiley, Hoboken, NJ. DOI: 10.1002/0471754323. 113, 114, 118, 133, 134, 135, 136, 137, 139, 141, 142, 149, 150, 165, 183

[44] Engheta, N. and Zsolkowskie, R., Ed., 2006, *Metamaterials: Physics and Engineering Explorations*, Wiley, Hoboken, NJ.

[45] Shekhar, P., Atkinson, J., and Jacob, Z., 2014, Hyperbolic metamaterials: Fundamentals and applications, *Nano Convergence*, 1(1):14. DOI: 10.1186/s40580-014-0014-6. 150

[46] Farnell, G. W., 1970, Properties of elastic surface waves, *Physical Acoustics*, 6:109–166. DOI: 10.1016/b978-0-12-395666-8.50017-8. 153, 154, 155, 156, 157, 158, 184

[47] Farnell, G. W., 1978, Types and properties of surface waves, In *Acoustic Surface Waves, Topics in Applied Physics, 24*, Springer-Verlag, Berlin and New York. DOI: 10.1007/3-540-08575-0_9.

[48] Rayleigh, L., 1881, *Philosophical Magazine 23*, p. 81. 153, 155

[49] Maradudin, A. A. and McGurn, A. R., 1989, Scattering of a surface-skimming bulk transverse wave by an elastic ridge, *Physical Review B39*, pp. 8732–8735. DOI: 10.1103/physrevb.39.8732. 155

[50] McGurn, A. R. and Maradudin, A. A., 1989, Localization of shear horizontal surface acoustic waves on a disordered surface, *Physical Review B39*, pp. 2125–2133. DOI: 10.1103/physrevb.39.2125. 155

[51] Maradudin, A. A., Ryan, P., and McGurn, A. R., 1988, Shear horizontal acoustic surface shape resonances, *Physical Review B38*, pp. 3068–3074. DOI: 10.1103/physrevb.38.3068. 153, 158

[52] McGurn, A. R., 2017, Kerr nonlinear layered photonic crystal coatings, *Proc. SPIE 10112*. DOI: 10.1117/12.2250040. 154, 163

[53] Le Rue, E. and Gabriel, P., 2009, *Principles of Surface-Enhanced Raman Spectroscopy: And Related Effects*, Elsevier, Amsterdam, Chapters 1, 2, 3, and 4. DOI: 10.1007/978-3-319-23992-7. 154, 164, 182

[54] Prochazka, M., 2016, *Surface-Enhanced Raman Spectroscopy: Bioanalytical, Biomolecular and Medical Applications*, Springer, Cham, Chapters 2 and 3. 154

[55] Stiles, P. L., Dieringer, J. A., Shah, N. C., and Van Duyne, R. P., 2008, Surface-enhanced Raman spectroscopy, *Annual Review of Analytical Chemistry*, 1:601–626. DOI: 10.1146/annurev.anchem.1.031207.112814.

[56] Fan, M., Andrade, G. F. S., and Brolo, A. G., 2011, A review on the fabrication of substrates for surface enhanced Raman spectroscopy and their applications in analytical chemistry, *Analytica Chimica Acta 693*, pp. 7–25. DOI: 10.1016/j.aca.2011.03.002. 164, 182

[57] Brevet, P.-F., 1997, *Surface Second Harmonic Generation*, Presses Polytechniques et Universitaires Romandes, Paris, Chapters 9 and 10. 164, 182

[58] Lee, J. L., Ty, C., Chen, P.-Y., Lu, F., Demmerie, F., Boehm, G., Amann, M.-C., Alu, A., and Belkin, M. A., 2014, Giant nonlinear response from plasmonic metasurfaces coupled to intersubband transition, *Nature 511*, pp. 65–69. DOI: 10.1364/cleo_qels.2014.fth4k.1.

[59] Wolf, O., Campione, S., Benz, A., Liu, S., Luk, T. S., Kadlec, E. A., Shaner, E. A., Klem, J., Sinclaire, M. B., and Brener, I., 2015, Phased-array sources based on nonlinear metamaterial nanocavities, *Nature Communications 6*, p. 7667. DOI: 10.1038/ncomms8667.

[60] Campione, S., Benz, A., Sinclaire, M. B., Capolino, F., and Brener, I., 2014, Second harmonic generation from metamaterials strongly coupled to intersubband transitions in quantum wells, *Applied Physics Letters 104*, p. 131104. DOI: 10.1063/1.4870072.

[61] Shen, Y. R., 1985, Optical second harmonic generation for surface studies, in *The Structure of Surfaces*, Eds. Van Hove, A., and Tong, S. Y., Springer, Berlin, pp. 77–83. DOI: 10.1007/978-3-642-82493-7_14.

[62] Shen, Y. R., 1989, Surface properties probed by second harmonic and sum-frequency generation, *Nature 337*, pp. 519–525. DOI: 10.1038/337519a0.

[63] Ren, M.-L., Liu, W., and Agarwal, R., 2014, Enhanced second-harmonic generation from metal-integrated semiconductor nanowires via highly confined whispering gallery modes, *Nature Communications 5*, p. 5432. DOI: 10.1038/ncomms6432. 164, 182

[64] Pohl, D. W. and Courjon, D., 1993, *Near Field Optics*, Springer, Berlin, pp. 147–286. DOI: 10.1007/978-94-011-1978-8. 164, 165, 183

[65] Paesler, M. A. and Moyer, P. J., 1996, *Near-Field Optics: Theory, Instrumentation, and Applications*, Wiley-Interscience, New York, Chapters 1–5. DOI: 10.1063/1.882005.

[66] Courjon, D., 2003, *Near-Field Microscopy and Near-Field Optics*, World Scientific Publishing Company, Singapore, Chapters 1, 2, 3, and 4. DOI: 10.1142/p220.

[67] Synge, E. H., 1928, Suggested method for extending microscopic resolution into the ultra-microscopic region, *Philosophical Magazine 6*, pp. 356–362. DOI: 10.1080/14786440808564615. 164

[68] Lui, H. and Lalanne, P., 2008, Microscopic theory of the extraordinary optical transmission, *Nature 452*, pp. 728–731. DOI: 10.1038/nature06762. 164

[69] Ebbensens, T. W., Lezed, H. J., Ghaemi, H. F., Thio, T., and Wolff, P. A., 1998, Extraordinary optical transmission through sub-wavelength hole arrays, *Nature 391*, pp. 667–669. DOI: 10.1038/35570. 164, 183

[70] Smolyaninov, I. I., 2018, *Hyperbolic Metamaterials*, Morgan & Claypool, San Rafael, Chapters 1, 2, and 3. DOI: 10.1088/978-1-6817-4565-7. 163, 164, 165, 183

[71] Shekhar, P., Atkinson, and Jacob, Z., 2014, Hyperbolic metamaterials: Fundamentals and applications, *Nano Convergence*, 1:14. DOI: 10.1186/s40580-014-0014-6. 164, 165, 183

[72] Grey, R. M., 2011, *Entropy and Information Theory*, Springer, New York, Chapter 3. DOI: 10.1007/978-1-4419-7970-4. 165, 169, 170

[73] Pierce, J. R., 1980, *An Introduction to Information Theory: Symbols, Signals and Noise*, Dover Publications, Inc., New York, Chapter 5. 165, 169, 170

[74] Michalowicz, J. V., Nichols, J. M., and Bucgoltz, F., 2013, *Handbook on Differential Entropy*, CRC Press, Boca Raton, FL, Chapter 4. DOI: 10.1201/b15991. 169, 171, 172

[75] Luo, L., Wang, J., Zhang, L., and Zhang, S., 2016, The differential entropy of the joint distribution of eigenvalues of random density matrices, *Entropy 18*, pp. 342–365. DOI: 10.3390/e18090342. 169, 170, 171, 172

[76] Maiers, S. A., 2007, Excitation of surface plasmon polaritons at planar interfaces, in *Plasmonics: Fundametals and Applications*, Springer, New York, pp. 39–52. DOI: 10.1007/0-387-37825-1_3. 183

[77] Stockman, M. I., 2013, Spaser, plasmonic amplification, and loss compensation, in *Active Plasmonics and Tuneable Plasmonic Metamaterials*, Eds. A. V. Zayats and S. A. Maiers, John Wiley & Sons, NJ, pp. 1–39. DOI: 10.1002/9781118634394.ch1. 183

[78] Bozhevolnyi, S. I., 2009, *Plasmonic Nanoguides and Circuits*, Pan Stanford Publishing, Danvers, MA. DOI: 10.1142/9789814241335. 183

[79] Huang, X., El-Sayed, I., El-Sayed, M., Chen, P., and Oyelere, A., 2007, Molecular cancer targeting and diagnosis using plasmonic gold nanoparticles, *Clinical Cancer Research 13*, 19:A22. 183

[80] Li, E.-P. and Chu, H.-S., 2014, *Plasmonic Nanoelectronics and Sensing*, Cambridge University Press, Cambridge. DOI: 10.1017/cbo9781139208802. 184

[81] Jang, Y. H., Jang, Y. J., Kim, S., Quan, L. N., Chung, K., and Kim, D. H., 2016, Plasmonic solar cells: From rational design to mechanical overview, *Chemical Reviews 116*, 2:14982–15034. DOI: 10.1021/acs.chemrev.6b00302. 184

Author's Biography

ARTHUR R. MCGURN

Professor Emeritus Arthur R. McGurn, CPhys, FInstP, is a Fellow of the Institute of Physics, a Fellow of the American Physical Society, a Fellow of the Optical Society of America, a Fellow of the Electromagnetics Academy, and an Outstanding Referee for the journals of the American Physical Society. He received a Ph.D. in Physics in 1975 from the University of California, Santa Barbara, followed by postdoctoral studies at Temple University, Michigan State University, and George Washington University (NASA Langley Research Center). The continuing research interests of Prof. McGurn have included works in the theory of: magnetism in disorder materials; electron conductivity; the properties of phonons; ferroelectrics and their nonlinear dynamics; Anderson localization; amorphous materials; the scattering of light from disordered media and rough surfaces; the properties of speckle correlations of light; quantum optics; nonlinear optics; the dynamical properties of nonlinear systems; photonic crystals; and meta-materials. He has over 150 publications spread amongst these various topics. Since 1981 he has taught physics for 38 years at Western Michigan University where he is currently a Professor Emeritus of Physics and a WMU Distinguished Faculty Scholar. A number of Ph.D. students have graduated from Western Michigan University under his supervision. He has previously published two books: *Nonlinear Optics of Photonic Crystals and Meta-Materials* (2015) and *Nanophotonics* (2018).

Printed in the United States
by Baker & Taylor Publisher Services